LIFE SCIENCE LIBRARY

THE SCIENTIST

TIME-LIFE BOOKS

Life World Library

Life Nature Library

Life Science Library

The Life History of the United States

Life Pictorial Atlas of the World

The Epic of Man

The Wonders of Life on Earth

The World We Live In

The World's Great Religions

The Life Book of Christmas

Life's Picture History of Western Man

The Life Treasury of American Folklore

America's Arts and Skills

300 Years of American Painting

The Second World War

Life's Picture History of World War II

Picture Cook Book

Life Guide to Paris

Time Reading Program

LIFE SCIENCE LIBRARY

CONSULTING EDITORS
René Dubos
Henry Margenau
C. P. Snow

THE SCIENTIST

by Henry Margenau, David Bergamini
and the Editors of LIFE

TIME INCORPORATED, NEW YORK

ABOUT THIS BOUK

TODAY THE FRUITS OF SCIENCE are evident everywhere, but the man responsible for them has remained, in large part, an enigma. Both by tradition and by preference, the scientist tends to be self-effacing. Yet never in history has there been a more urgent need to understand who he is and how he works. This book scrutinizes the scientist as a human being and as a modern legend, as a thinker and a doer, and as a powerful new force in 20th Century life.

In this volume, text chapters alternate with picture essays, and each may be read independently of the other, but the essays are designed primarily to supplement the text. For example, Chapter 3, "The Scientific Method," describes the scientist's unique style of inquiry; "The Pursuit of Omega Minus," the essay which follows, shows how the application of this method produced one vital scientific discovery.

THE AUTHORS

HENRY MARGENAU has played a variety of distinguished roles in the world of science: physicist, teacher, adviser to government and industry, editor and writer. He is also a permanent consulting editor of the LIFE Science Library (below).

DAVID BERGAMINI is a freelance author who specializes in the field of science. His previous books include *The Universe* and *The Land and Wildlife of Australia* for the LIFE Nature Library, and *Mathematics* for the LIFE Science Library.

THE CONSULTING EDITORS

RENE DUBOS, a member and professor of The Rockefeller Institute, is a microbiologist and experimental pathologist world-famous for his pioneering in antibiotics, including the discovery of tyrothricin. He has written, among other books, *Mirage of Health* and *The Dreams of Reason*.

HENRY MARGENAU is Eugene Higgins Professor of Physics and Natural Philosophy at Yale, an editor of the *American Journal of Science* and a notable contributor to spectroscopy and nuclear physics. His books include *Open Vistas* and *The Nature of Physical Reality*.

C. P. SNOW has won an international audience for his novels, including *The New Men*, *The Affair* and *Corridors of Power*, which explore the effects of science on today's society. Trained as a physicist, he directed recruitment of scientific personnel for Britain's Ministry of Labour in World War II. He was knighted in 1957.

ON THE COVER

A 1962 Nobel laureate for his co-discovery of the DNA molecule, Dr. James D. Watson of Harvard personifies today's scientist in his role as theoretician—a creator of conceptual patterns that reflect the underlying order in nature. The symbols on the back cover represent the scientist's basic task: to piece together the fragments of nature's infinite puzzles.

The Scientist © 1964 Time Inc. All rights reserved.
Published simultaneously in Canada. Library of Congress catalogue card number 64-8795.
School and library distribution by Silver Burdett Company.

CONTENTS

		PAGE
	INTRODUCTION	7
1	**HERO—AND HUMAN BEING**	8
	Picture Essay: A Landscape of Poetic Vision 16	
2	**PROFILE OF A NEW ELITE**	28
	Picture Essay: The Instruments of Conquest 36	
3	**THE SCIENTIFIC METHOD**	50
	Picture Essay: The Pursuit of Omega Minus 62	
4	**AN EXPANDING REALM**	74
	Picture Essay: The Family Trees of Science 84	
5	**THE COMMUNICATIONS GULF**	102
	Picture Essay: The Voice of the Scientist 110	
6	**A BOOMING ESTABLISHMENT**	122
	Picture Essay: California, "the Science State" 130	
7	**THE BOUNTY OF TECHNOLOGY**	144
	Picture Essay: Old Guesses about the Future 152	
8	**THE IMPACT OF SCIENCE**	164
	Picture Essay: The Nobel Prize: Accolade for Greatness 174	
	APPENDIX	189
	A Gallery of Nobel Laureates	
	BIBLIOGRAPHY AND ACKNOWLEDGMENTS	196
	INDEX	197
	PICTURE CREDITS	200

TIME-LIFE BOOKS
EDITOR
Norman P. Ross
TEXT DIRECTOR   ART DIRECTOR
William Jay Gold   Edward A. Hamilton
CHIEF OF RESEARCH
Beatrice T. Dobie
Assistant Text Director: Jerry Korn
Assistant Chief of Research: Monica O. Horne

PUBLISHER
Rhett Austell
General Manager: Joseph C. Hazen Jr.
Business Manager: John D. McSweeney
Circulation Manager: Joan D. Lanning

LIFE MAGAZINE
EDITOR: Edward K. Thompson
MANAGING EDITOR: George P. Hunt
PUBLISHER: Jerome S. Hardy

LIFE SCIENCE LIBRARY
SERIES EDITOR: Robert Claiborne
Editorial staff for *The Scientist:*
Associate Editor: Robert G. Mason
Text Editor: Diana Hirsh
Assistant Text Editor: Neal G. Stuart
Picture Editor: Sheila Osmundsen
Designer: Arnold C. Holeywell
Associate Designer: Edwin Taylor
Staff Writers: Tom Alexander, Jonathan Kastner, Harvey B. Loomis, Charles Osborne, Gerald Simons, Edmund White
Chief Researcher: Thelma C. Stevens
Researchers: Sarah Bennett, Robert W. Bone, Mollie Cooper, Owen Fang, Ann Ferebee, Jane M. Furth, Penny Grist, Susanna Seymour, Patricia Tolles

EDITORIAL PRODUCTION
Art Associate: Robert L. Young
Art Assistants: James D. Smith, Patricia Byrne, Charles Mikolaycak, Douglas B. Graham
Picture Researchers: Margaret K. Goldsmith, Susan Boyle
Copy Staff: Marian Gordon Goldman, Suzanne Seixas, Dolores A. Littles, Clio Vias

The text for the chapters of this book was written by Henry Margenau and David Bergamini, for the picture essays by the editorial staff. The following individuals and departments of Time Incorporated were helpful in the production of the book: Larry Burrows, Alfred Eisenstaedt and Art Rickerby, LIFE staff photographers; George Karas, Chief of the LIFE Photographic Laboratory; Margaret Sargent, LIFE film editor; Doris O'Neil, Chief of the LIFE Picture Library; Richard M. Clurman, Chief of the TIME-LIFE News Service; and Content Peckham, Chief of the Time Inc. Bureau of Editorial Reference.

INTRODUCTION

ONE OF THE MOST ASTONISHING FEATURES of the evolutionary development which led to Homo sapiens was the vast *latent* capacity with which the new species was endowed, far beyond what could have been the immediate needs. We were given the ability to learn new things not only by trial and error but also by logical processes. Moreover, we were provided with an enormous potential talent for expressing ideas, both practical and abstract, in readily communicable form. Presumably the gifts of logical analysis and communication of complex ideas first developed because of the extraordinary advantage they gave our species in the various day-to-day competitions of life, both inside and outside the human community. However, they also empowered us to build a bridge from the earth to the stars—as long as we were willing to dream, and to act upon the dream.

It is somewhat disquieting to speculate on the fact that even 50,000 years ago, in the early Stone Age, the human family contained individuals with innate capacities for reasoning and self-expression approaching those of a Shakespeare, a Beethoven or an Einstein. Only the common fund of knowledge and experience accumulated by intervening generations allows us today to make fuller use of our genetic inheritance. To that fund the scientist, through the practice of his calling, has been a major contributor.

What we today call science is a relatively new acquisition in the human journey. Its crucially important philosophical aspects, which transcend the purely practical and serve to broaden our minds and vision so enormously, were first appreciated in an explicit way by the classical Greeks about 600 B.C. But the Greeks and later scholars of the classical world never achieved in a uniform way the tight interrelation between theory and experiment which is one of the most marked characteristics of present-day science. Today the flow from relatively pure or abstract scientific discovery to practical application is such that science, which started out primarily as an adventure of the mind, is now becoming one of the principal pillars of our way of life.

By the same token, the man of science is becoming one of the prime movers in our society. Increasingly we want—and need—to know all we can about him. This LIFE book, part of a notable series, brings the world of the scientist to the view of the nonscientist. It describes the atmosphere in which he lives and works. It sheds light on the way he thinks. Above all, it reminds us that, regardless of the diversity of nationalities, personalities and specialties, scientists everywhere—intentionally or not—are bound together by a common purpose: the advancement of human capabilities.

—FREDERICK SEITZ
President, National Academy of Sciences

1
Hero—
and Human Being

HISTORY ON A SCRATCH PAD
This seemingly casual and meaningless jumble of jottings is a major document of science because of the "P-violated" notation *(right center)*. The Nobel Prize-winning work of physicists Tsung Dao Lee and Chen Ning Yang, it upset the law of conservation of parity (P), which had assumed the symmetry of the universe, and suggested instead that space has a kind of twist.

ONE MUGGY DAY in July 1939, two eminent scientists in an automobile found themselves lost in the wilds of Long Island. They had come on a mission so disconcertingly important, so melodramatic and irregular, that they had neglected to make sure of their directions.

"Perhaps I misunderstood on the telephone," one of them ventured. "I thought he said Patchogue."

"Could it have been Cutchogue?" asked the other, a trifle irked.

Some time later they pulled up on a street in Cutchogue and inquired the way to "Dr. Moore's cabin," but to no avail. As they drove around, feeling more and more frustrated, one of them said:

"Maybe fate never intended it. Let's go home."

"Wait," suggested the other. "What if we simply inquire where Einstein is staying?" They stopped at the curb beside a small, sunburned boy of about seven, and asked if he knew Professor Einstein.

"Sure," he said. "Want me to take you to him?"

And so it came about, the story goes, that Eugene Wigner and Leo Szilard, two Hungarian refugee physicists, finally reached Albert Einstein on an afternoon two months before the start of World War II and persuaded him to write a letter to President Roosevelt urging quick action to offset possible Nazi progress on an A-bomb.

It is altogether appropriate that a child—a messenger, as it were, of the future—should have served fate to usher in the atomic age. In the public mind the atom has come to symbolize the power and glory of today's scientist. As it happened, at about the time that Wigner and Szilard crossed the threshold of Einstein's vacation retreat, the number of scientists in the world passed the million mark. Big science was born. Before long, the era of the atom was to merge with the era of the computer, of space travel and of the chemical analysis of heredity. The scientist, by the hand of an innocent, was led out of his quiet laboratory and classroom into the din and glare of launch pads, Congressional hearings and councils of state.

The new age has barely dawned, and already we live in a world of satellites and genetic chemistry which, that summer of 1939, would have seemed pure science fiction. In his professorial, often inarticulate fashion the scientist has come to shape the policies of nations. Within a few decades, through technology, he has radically transformed the trappings of personal life, including our clothes, food, entertainment and bank balances. From foundations laid centuries ago, he has built upward with breathtaking rapidity. The roof of his structure is still out of sight, riding a cyclone on the way to Oz.

Who is this master builder, the scientist? He is, to begin with, a human being subject to all the strengths and weaknesses of his fellowmen. At the same time the very nature of his lifework makes him a breed apart. Beyond his own fraternity—now numbering some six million around the world—there are those who fear and suspect him; those who admire and reverence him; and few who know him well. His achievements are often celebrated, the man behind them seldom.

There is much for the layman to know about the scientist: the discernible patterns in his personality, the unique style of inquiry which stamps his work, the steady spread of his domain, his communication and organization problems in the age of bigness he himself has helped bring on, and the ways in which he leaves his mark upon every one of us, every day, in things of the flesh, the mind and the spirit alike. Each of these matters will be taken up in subsequent chapters of this book.

We start with a look at the man in the round, by way of a collection of stories. These next pages show the scientist tinkering and theorizing, joking and in jeopardy, dreaming absently and thinking at full tilt, writing cosmic equations and marveling at butterfly wings. Behind such moments lie gaiety and grief for the scientist, certainty and bafflement, triumph and defeat. Sensing this, we may begin to take the measure of the man in his dual role: as a member of a special breed and as a member of the human race.

❊   ❊   ❊   ❊   ❊

"Now, don't be so modest, Professor. I'm sure you've got something up your sleeve that will blow us all to bits."

© 1955
The New Yorker Magazine, Inc.

In the august halls of the Institute for Advanced Study at Princeton, one day in the late 1940s, Dr. Walter Stewart, an economist on the staff, stood and watched a number of young graduate students in physics as they came bursting out of a seminar. They were, he was later to recall, "beyond all doubt the noisiest, rowdiest, most active and most intellectually alert" of all the Institute's budding talents. That day Stewart stopped one of them as he charged past and asked: "How did it go?" "Wonderful!" came the reply. "Everything we knew about physics last week isn't true!"

❊   ❊   ❊   ❊   ❊

The office of the great astronomer Walter Baade would have caused any self-respecting corporation executive to shudder. The room was about the size of a walk-in closet, its one window overlooked a Pasadena parking lot, and the desk lay buried beneath a foot-high debris of scribbled memos and photographic plates. Baade, one of the leading 20th Century explorers of our universe, was, by his own confession, "lazy." He saw no point, when there was so much to be discovered in a lifetime, in taking time to write out his ideas in formal fashion. Yet amid the clutter on his desk, according to an admiring colleague, lay "half the secrets of the cosmos, crumpled up in small smudged scraps of paper."

After Baade's death in 1960, friends, digging through his notes in expectation of coming upon some fresh cosmic theories, found themselves considerably discomfited. Such is the pace of modern astronomical discovery that they unearthed little new knowledge which had not, by then, been independently brought to light by others in the profession.

*   *   *   *   *

In science, the race is to the young, the swift—and the skeptical. Laymen may exalt the complicated new gadgets of the craft; not so the men who daily deal with them. In a basement room at the Argonne National Laboratories in Chicago, hard by an electronic computer, is a glass case bearing the inscription, "In case of emergency, break glass." Displayed inside is an abacus.

*   *   *   *   *

The scientific tradition of ingenuity never dies. The Nobel Prize-winning geneticist Thomas Hunt Morgan once hit on the idea that acidity in sea water may increase the fertility of certain creatures of the deep. Having no acid on hand, he went out to the nearest grocery store, bought a lemon, squeezed its juice into his aquarium and thereby made scientific history by showing that only a small change in chemical environment can greatly affect the fundamental life process of fertilization.

*   *   *   *   *

In 1952, before there were earth satellites to help investigate the upper atmosphere, James A. Van Allen was trying to make do with small rockets launched at 70,000 feet from Navy weather balloons. The man who was later to discover the existence of radiation belts in outer space was having a hard time of it: the rockets repeatedly failed to fire. Van Allen decided that the extreme cold in the upper air was affecting the rockets' clockwork. He rooted around in the galley of his balloon-launching vessel, found some cans of orange juice, heated them up and clustered them around the firing mechanism of his next rocket package. Thus warmed, the rockets functioned perfectly.

*   *   *   *   *

Alastair Pilkington, the technical director of a large British glass-manufacturing firm, devised a revolutionary new method by which perfect plate glass could be made without recourse to the grinding process. Pilkington's method was to float molten glass in a continuous ribbon off molten tin; he got the idea while watching soapsuds on dishwater as he helped his wife with the dinner dishes. The British meteorologist Sir Geoffrey Taylor designed the lightweight anchors used in the gigantic artificial harbors at Normandy on D-Day; he got the idea for their basic shape while contemplating an old plow he found lying in a ditch.

*   *   *   *   *

What will inspire an idea, no scientist is wise enough to foresee; what will come of it, no scientist is rash enough to predict. In 1960 Richard P. Feynman, professor of theoretical physics at the California Institute of Technology, became enamored of the possibilities of microminiaturiza-

"That's the trouble with scientist husbands. It's always 'work in progress' up here."

tion. In a public lecture, he offered to give $1,000 of his own money to anyone who would build him an electric motor no more than a quarter millionth of a cubic inch in volume. Feynman is noted as a prankster; at Los Alamos, during World War II, he used to crack safes just for the pleasure of leaving "guess who" notes for security officers. This time, however, he was in dead earnest. If anyone ever took him up on his offer, he figured, important new principles might be evolved which would shed light on such microminiaturized mechanisms as those which work in living cells.

Over the ensuing months Feynman was beset by inventors of motors which, though flea-sized, far exceeded his specifications. Then one day, on campus, he was approached by a Pasadena engineer named William H. McLellan, carrying a package the size of a shoe box. Somewhat impatiently, Feynman watched McLellan rip it open; to his astonishment he saw only a microscope. Peering through the eyepiece, he discovered a synchronous electric motor no larger than a dust mote. Ingeniously fashioned with the aid of a microdrill press and a watchmaker's lathe, it ran in the same manner as motors that weigh tons.

Feynman promptly paid up, but cautiously withdrew a second $1,000 offer for anyone who managed to reduce the contents of a book page to 1/25,000 the original size. Somewhat abashed, he explained: "In the meantime I've gotten married and bought a house."

"Say, I think I see where we went off. Isn't eight times seven fifty-six?"

© 1954
The New Yorker Magazine, Inc.

✻ ✻ ✻ ✻ ✻

Far-out ideas also fascinate the scientists of AMSOC, but generally just for laughs. Founded as a typically gentle scientific spoof, AMSOC stands for American Miscellaneous Society. Its members are some 50 men who enjoy partying together, usually in Washington. Its imaginary ex-officio branches include committees on calamitology and trivialogy; its awards, if any, are given for "dynamic incompetence," "distinguished obscurity" and "unwarranted assumptions." For a time the Society's favorite wild idea was a proposal to tow Antarctic icebergs to the Pacific Coast so that they could be shipped inland and melted to irrigate the desert. Then AMSOC member Walter Munk, the oceanographer, had his brainstorm.

Munk's idea, which he sprang on his fellow members at a wine breakfast at his California home in 1957, was no more, no less, than to drill a hole right through the crust of the earth, 20 or more miles down from the surface of the land, or some three miles down from the ocean bottom. Such a hole, Munk asserted, would help to clarify the processes by which the earth, its oceans and continents came into being. It would reveal billions of years of marine evolution. It would penetrate all the way to where the earth's crust abruptly gives way to a second layer of material, the mantle, a region seismologists call the "Mohorovičić discontinuity."

The more Munk expounded on his idea, the less fantastic it seemed.

Through AMSOC's efforts, it was proved feasible by a 600-foot test bore under 11,700 feet of water off California. The achievement was an astonishing one; never before had there been drilling at a level below 200 feet of water. Later the project, dubbed Mohole, was turned over to the National Science Foundation and given the status of a multi-million-dollar government-sponsored enterprise operated under contract by a Texas construction firm.

*****

The scientist's celebrated absent-mindedness stems primarily from his preoccupation with the problem that seems most important at the moment. In 1933 Caltech's senior seismologist, Beno Gutenberg, received a visit from Einstein, who wanted to know something of Gutenberg's specialty. The two strolled around the campus while Gutenberg explained the science of earthquakes. Suddenly an excited colleague broke in on them. They looked around to see people rushing from buildings and the earth heaving under their feet. "We had become so involved in seismology," recalls Gutenberg, "that we hadn't noticed the famous Los Angeles earthquake, the biggest I had ever experienced, taking place around us."

*****

In pursuit of a discovery, the dedicated scientist will do almost anything. The British physicist Lord Cherwell learned to fly in three weeks during World War I in order to put a plane into a dangerous spiral dive and prove, successfully, his solution to a problem in aerodynamics. The Austrian zoologist Konrad Lorenz, to demonstrate that ducklings can be "imprinted" with almost any sort of "mother image," used to waddle about his backyard in a crouch, quacking loudly and followed by an Indian file of obedient chicks. One day he looked back to see how they were behaving and noticed a row of horrified human faces peering over the fence. "Surely," he thought, "it is obvious that I am imprinting ducklings." Then he realized that his charges were hidden from the spectators by the tall grass.

*****

The scientific sense of humor tends to be cerebral and most intramural. For the marathon calculations required to build the hydrogen bomb, the formidable mathematical genius John Von Neumann devised a machine which he christened "Mathematical Analyzer, Numerical Integrator and Computer." The arithmetizing monster was duly delivered and proved its worth. Then its users discovered among its other virtues a built-in gag: when alphabetized, its name shrank to MANIAC.

*****

George Gamow, a Russian physicist who fled to the West in 1933, collaborated some years later with one of his graduate students, Ralph

Alpher, on a theory about the origin of the chemical elements in the universe. They examined the possibility that all the various kinds of atoms could have been cooked up out of elementary particles in the first few seconds of the universe, when it was just beginning to expand from an initial state of extreme density. The idea worked out so well mathematically that they decided to report on it via the columns of the renowned *Physical Review*. For a work of such scope, however, they felt the need of a grand-sounding authorship. And so they added the name of the nuclear physicist Hans Bethe to their own, and came up with *The Origin of Chemical Elements* by Alpher, Bethe and Gamow. To this day their presentation remains one of the ABCs of cosmological thought, and one of the alpha-beta-gammas in any scientist's repertoire of jokes.

※ ※ ※ ※ ※

Even in his more physical forms of recreation, the scientist is unlikely to leave science far behind. Luis Alvarez, professor of physics at the University of California in Berkeley, is a golfer. Hoping to improve his game, he invented a stroboscopic analyzer, a device for studying the successive phases of motion by the use of light. Attached to the golf club, it enabled Alvarez to scrutinize every stroke he made. He presented a model to the celebrated fellow golfer then occupying the White House, Dwight Eisenhower, and has since taken out a patent (No. 2,825,569).

※ ※ ※ ※ ※

George Beadle, the Nobel laureate geneticist, is a mountain climber. Before he took his present post as President of the University of Chicago, he taught at Caltech and, whenever opportunity afforded, pursued his hobby in the California hill country. On weekends when he could not get away, Beadle would take piton, pick and rope, and scale the elaborate Gothic and Romanesque buildings on the campus. For good measure he would carry along his camera and photograph the structural carvings from odd angles, later trying to stump his friends by challenging them to identify the sites.

※ ※ ※ ※ ※

Some scientists have an almost monkish disregard for worldly possessions. This was particularly true of George Washington Carver, the great Negro agricultural chemist. Southern peanut growers, who profited by his work, often sought to give him gifts, and were constantly disappointed because he put the presents away and forgot them. Finally one wealthy Georgia planter, Tom Huston, asked Carver if there was nothing he really wanted. "Oh yes, a diamond," said Carver. Though taken aback, Huston bought a fine stone and had it mounted in a platinum setting. Carver accepted the gift gratefully but was never seen wearing it. Piqued, Huston commissioned a mutual friend to find out why. With great pride Carver led the friend to the case containing his

© 1947
The New Yorker Magazine, Inc.

collection of mineralogical specimens. There shone the diamond, carefully nested and labeled.

* * * * *

The noted endocrinologist, John Cortelyou, President of De Paul University in Chicago, was elected secretary of a newly founded organization for Roman Catholic scientists. He promptly set about disbanding the group. Cortelyou, whose specialty is the study of endocrine glands in amphibian animals, explained his action thus: "There are no Catholic frogs."

* * * * *

The scientist often meets death with a rare grace, and with the same objectivity which has marked his life. In November 1926, Professor Francis Weld Peabody of the Harvard Medical School delivered a lecture on "The Care of the Patient," a masterpiece of clarity and compassion, destined to be reprinted many times. During the lecture, he appeared his usual vigorous 46-year-old self. His grooming was, as always, flawless, his words thoughtful, eloquent, meticulous. His listeners did not know that he, personally, had become a patient; that he had learned he had an inoperable cancer; that he knew this lecture might be his last. After his death, his boyhood friend William James said of his last months: "He knew as much as any one about the nature of his disorder, and every symptom that he experienced carried, for him, the fullest medical significance ... he wrote his last hospital report on the day before he died."

© 1955
The New Yorker Magazine, Inc.

* * * * *

To know nature is the scientist's consuming aim, and in pursuing it he embraces all of nature within his affectionate regard. As a youth, Robert Goddard, the father of modern rocketry, began dreaming of interplanetary travel while perched amid the branches of a cherry tree which overlooked the vistas of his native Massachusetts countryside. When the 1938 hurricane swept New England, Goddard's property, now rented, lay in its path. On hearing that the orchard of his boyhood had been devastated, Goddard wrote in his private journal: "Cherry tree down. Have to carry on alone."

* * * * *

A huge, sky-blue *Morpho* butterfly zigzagged down to light on a leaf at the foot of a tree. Nicholas Guppy, a young British botanist—a relative of the Guppy who gave his name to the fish—abruptly held up a machete, signaling for silence. The Carib Indians and the two white men with him—his companions on a trip into the tropical rain forest of Surinam in 1953—obligingly stood still. Guppy shambled toward the resting butterfly, knelt and slowly held out his hand. The butterfly quivered, crawled across a speck of sunlight and climbed into the limp, outstretched knuckles. Guppy rose gently and held up the docile insect.

"Isn't it beautiful," he murmured. "I believe I have some kind of chemical attraction in the smell of my hands. I've never seen anyone else who could pick them up this way. It's *Morpho adonis*, incidentally. Such a pale, ethereal blue. Now you see why it is anyone would become a biologist. Learning about names and classifications and the dreary sex life of plants is all incidental to the main enthusiasm, which is the love for living things."

He flicked his wrist and the great insect, changing course with every stroke of its six-inch wings, flapped upward toward the treetops.

❖   ❖   ❖   ❖   ❖

Could scientists speak with one voice, they would, perhaps, choose to have us think of them in the summing-up of the philosopher Herbert Spencer. "Think you," he wrote, "that a drop of water, which to the vulgar eye is but a drop of water, loses anything in the eye of the physicist who knows that its elements are held together by a force which, if suddenly liberated, would produce a flash of lightning? . . . Think you that the rounded rock marked with parallel scratches calls up as much poetry in an ignorant mind as in the mind of a geologist, who knows that over this rock a glacier slid a million years ago?

"The truth is, that those who have never entered upon scientific pursuits know not a tithe of the poetry by which they are surrounded."

A Landscape of Poetic Vision

The work of the scientist is based upon a conviction that nature is basically orderly. Evidence to support this faith can be seen with the naked eye—in the design of a honeycomb or a mollusk's shell—but scientists come upon order at every level of being. The physicist finds it in the arrangement of atoms on a needle's point *(opposite)*, the entomologist in the structure of a mosquito's eye, the crystallographer in the architecture of crystals. The scientist's primary interest in order is in the information it supplies: the orderly laws of what, why and when. And when he finds the order he seeks he often finds beauty as well. The landscape, microscopic or macroscopic, that engrosses the scientist has symmetry, grace and balance. It is a landscape, according to MIT professor Gyorgy Kepes, that can delight "the scientist's brain, the poet's heart, the painter's eye . . . that has both the character of information and the quality of poetic vision."

ON THE POINT OF A NEEDLE
The picture on the opposite page shows the atoms that form the point of a platinum needle. The photograph was taken by Dr. Erwin Müller, inventor of the field ion microscope, which gives scientists the most intimate look at the microcosm they have ever had. The needle's point has been magnified two million times and every spot of reflected light represents a single atom or a cluster of atoms.

Intricate Symmetry

Elaborate designs are often found in nature. Even very simple plants and animals, like the mushroom or the chiton (a kind of mollusk), show symmetry of an intricate order. Every tree, no matter how gnarled outwardly, reveals inside a pattern of circles that tells the story of its growth.

UNDERSIDE OF A MUSHROOM CAP

TOP OF CHITON SHELL

CROSS SECTION OF YELLOW-PINE TREE TRUNK

CROSS SECTION OF BEAN-ROOT NODULE MAGNIFIED 40 TIMES

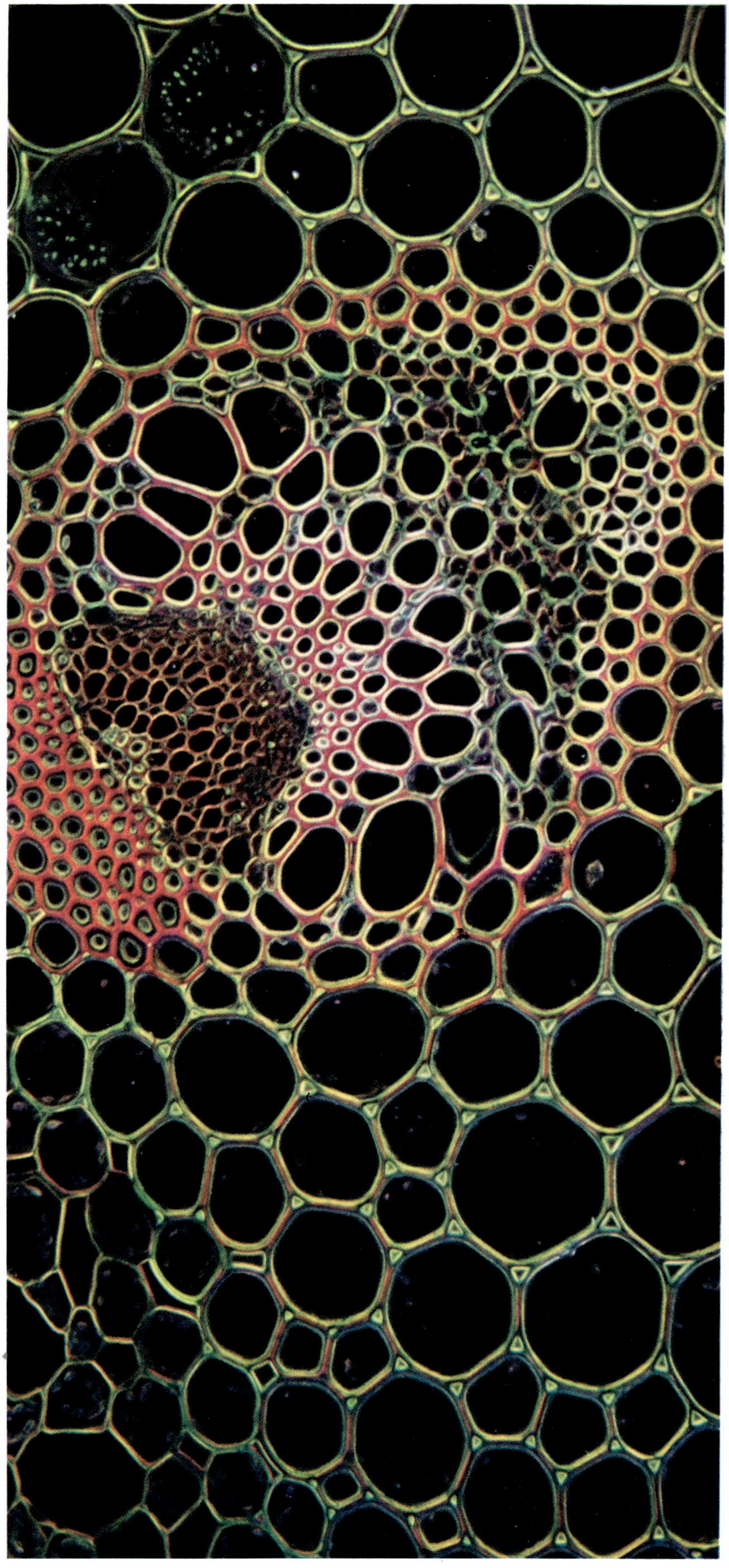

MOSQUITO'S EYE MAGNIFIED 160 TIMES

EEL SCALES MAGNIFIED 100 TIMES

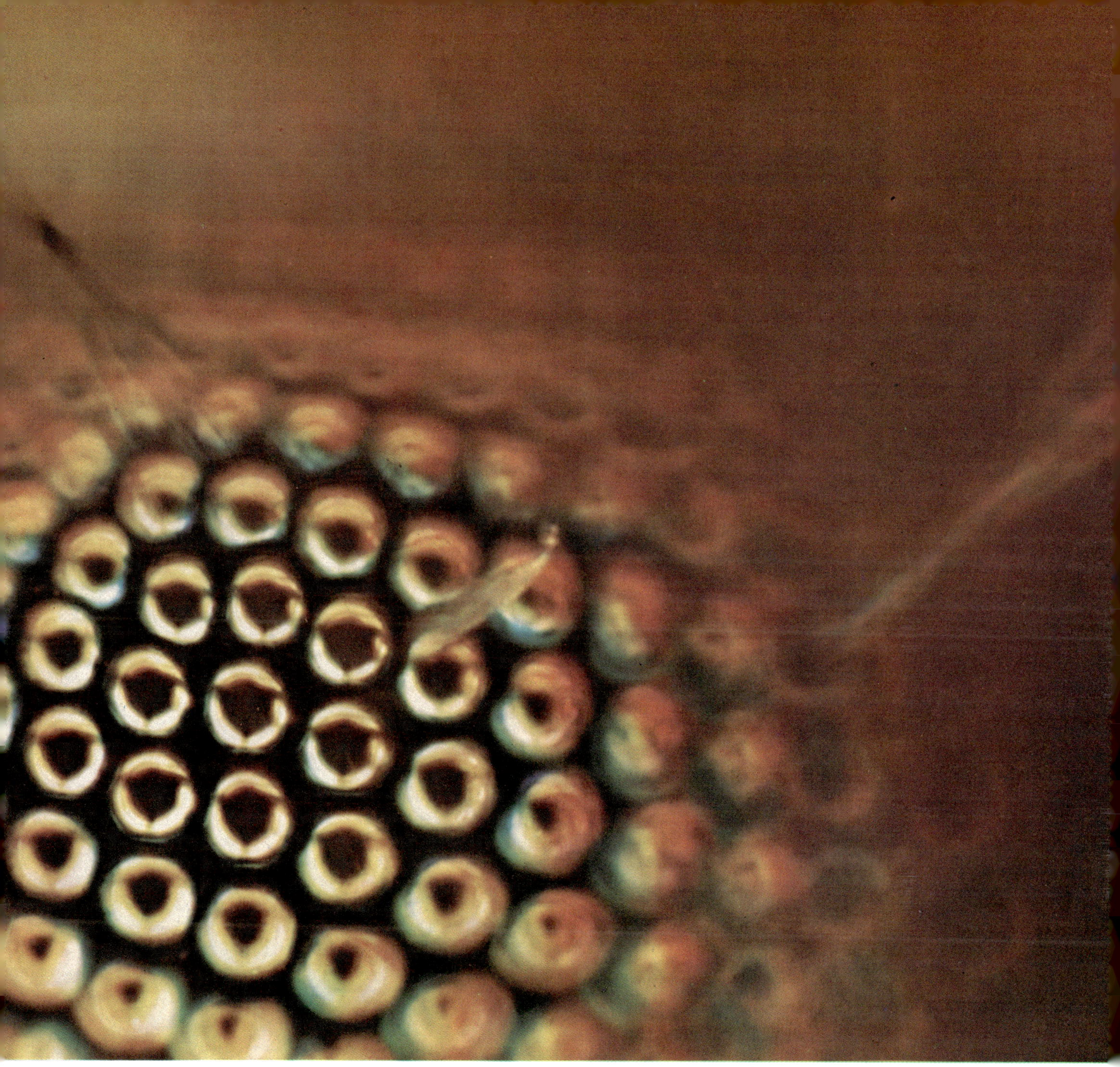

Shaped for Efficiency

Efficient natural shapes include the hexagon—the basic unit of the honeycomb, the mosquito's eye, the scales of an eel. The cells of a bean-root nodule grow in different shapes, yet their walls match up so well that the cells use every bit of space—and in addition produce an elegant design.

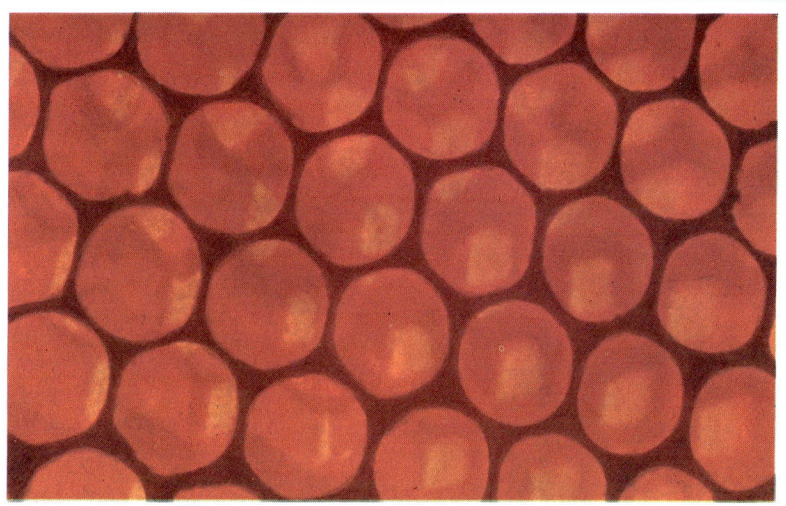

HONEYCOMB

The Dazzling Patterns of Light

The handsome design on the opposite page was made by metallurgists during a study of the effects of rolling friction on a pure copper crystal. The reflections of a beam of light, photographed with a magnification of 450 times, revealed the contours of a microscopic groove made by a tiny sapphire ball as it rolled across the face of the crystal. Even when light is meaningless, as in the design below, it falls into a pattern. Red, blue and green are the primary colors of light, and the only ones used by TV tubes. Their distribution on the tube creates the effect of different colors.

GROOVE IN THE FACE OF A COPPER CRYSTAL, MAGNIFIED 450 TIMES

COLOR-TV TUBE DURING ADJUSTMENT

IRON CRYSTAL, SLIGHTLY OXIDIZED, MAGNIFIED 10,000 TIMES

SILICON CRYSTAL MAGNIFIED 150 TIMES

IRON CRYSTAL, HEAVILY OXIDIZED, MAGNIFIED 10,000 TIMES

Crystal Building Blocks

Crystals are the building blocks of most metals and many other solids. The atoms or molecules of crystals build on one another with almost military precision, forming straight lines, edges and layers. The atoms of different elements line up at different angles. As the pictures on these pages show, silicon and iron atoms line up at right angles. The picture of a silicon crystal at left clearly discloses its cubic architecture. The smaller picture on the opposite page is of an iron crystal after brief exposure to oxidizing agents. Surface atoms oxidized in straight lines and right angles. The picture above shows an oxidized iron crystal whose corroded atoms have been removed, revealing the crystal's inner structure.

25

The Face of Disorder

Despite its fundamental order, nature often presents a disorderly face to the world. The internal order of an organization is always subject to intervention by external forces. The growth of a tree may be stunted by drought or fire, a light wave distorted by heat, a honeycomb ripped apart by a bear.

Crystals, for all their geometric elegance, are also subject to intervention. The crystalline mass shown at right was formed by many copper sulfate and calcium sulfate crystals growing together. Each had the potential to grow much larger, but was thwarted during its development by other crystals growing all around it. The result was a disorganized mass.

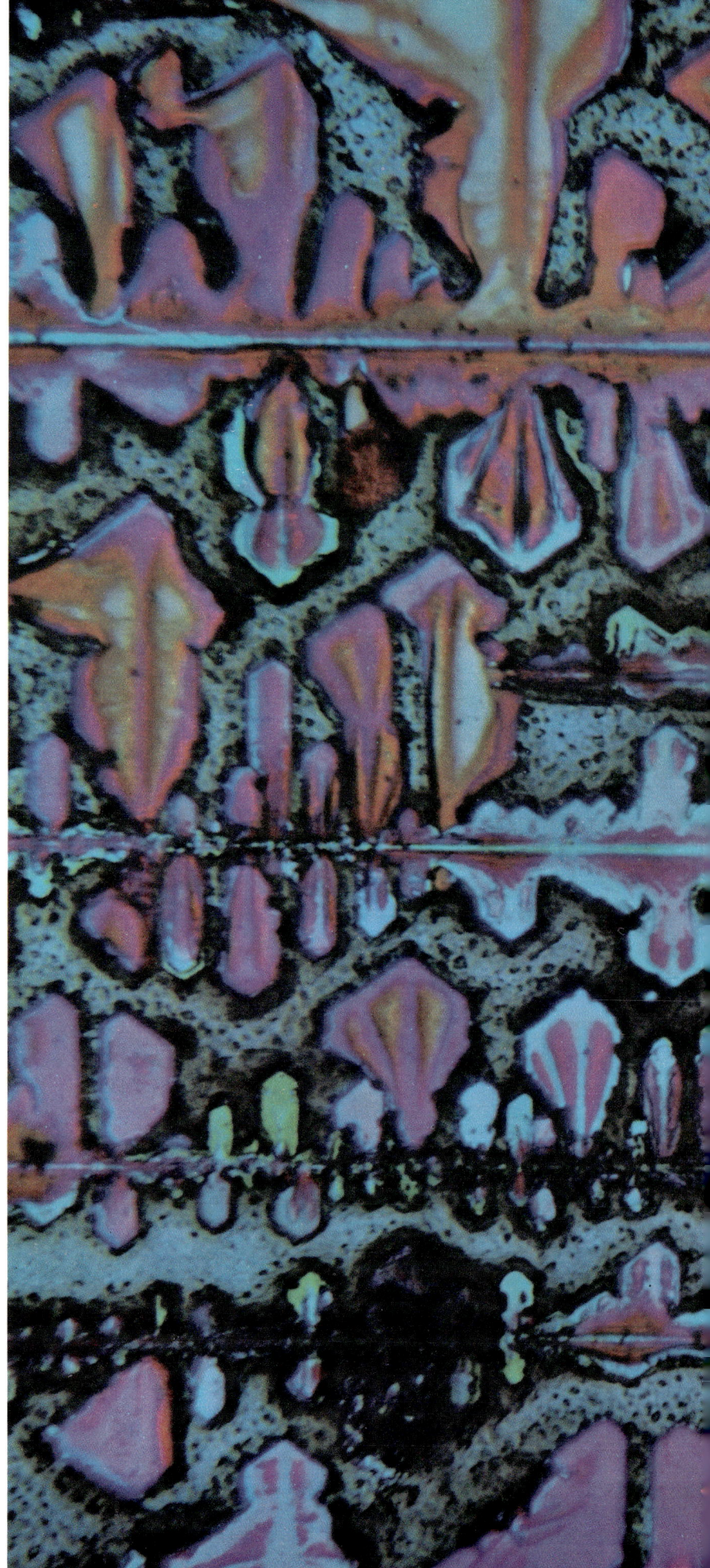

COPPER SULFATE AND CALCIUM SULFATE CRYSTALS

2
Profile of a New Elite

AFTER HIS DEEDS OF RECENT DECADES the scientist has come to be acclaimed as the mainspring of our complex age. Virtually everyone concedes his importance. Yet surprisingly few of us can define and clearly characterize him. Whether as a personality, a professional species, a public image or a social phenomenon, the scientist defies easy typecasting. In his various guises he may appear as an august scholar, a remote ascetic, a bright-eyed visionary or a sweat-soaked mechanic. While individually he is assumed to have the usual human frailties, collectively he is often seen looming larger than life: at one moment a kindly professorial god strewing comforts and plenties, and at the next a terrible blacksmith of atomic artillery.

The scientist poses an enigma to laymen partly because he is still a relative newcomer to their world. Until World War II, he moved mainly in academic circles. Even there his identity as a separate breed had been established for barely a century. Implausible as it may now seem, the very word "scientist" was coined only in 1840. Its inventor, a Cambridge historian and philosopher, William Whewell, introduced it as follows: "We need very much a name to describe a cultivator of science in general. I should incline to call him a scientist."

Cultivators of particular sciences enjoyed their special labels long before Whewell thought to assemble them under one tent. Astronomers were first mentioned in written English before 1400, mathematicians a few decades later. The chemist as a "distiller of waters" began to be set apart from the alchemist in the 1500s. The words "zoologist," to describe a student of animals, and "botanist," for a student of plants, were devised in the 1600s. They were not united under the more encompassing title of "biologist" until two centuries later. The "geologist" appeared in the 1700s. So did "psychologist," to differentiate a physician "of the soul" from the ordinary physician of the body. The term "physicist" was originated in 1840 by the same Whewell who hit upon "scientist."

The merging of all such specialists under the single banner of scientist has created a professional category so vast that the dictionaries find it almost impossible to define. Most fall back upon "man of science," or "one learned in science." Science, in turn, is often defined simply as "knowledge." Sometimes it is pinned down a bit further as knowledge "obtained by study and practice."

The Latin word *sciens* does, indeed, mean "knowing." In French *la science* is still a term for all types of learning, and in German *die Wissenschaft*, "knowledge," or the "art of knowing," is often used interchangeably with "science." Practitioners of science, however, have long believed that they pursue a kind of knowledge which differs from all others: one which is built entirely out of the brass tacks of fact and logic; which does not depend on historical report, majority opinion, fashion or taste;

THE HOME LIFE OF A SCIENTIST
Contrary to the image of the scientist as a loner too bemused to bother with his family, zoology professor Richard Huling of Ohio University regularly shares the joys of science with his four sons. From lending a hand in his home laboratory several years ago *(left)*, they have graduated to helping ready his research room on the university's campus for experiments.

and which can be demonstrated at any time and place to any willing human being with alert senses and a brain.

What most distinguishes scientific from other knowledge is the method by which it is created, a systematic extension of good sense and sound skepticism called the scientific method. The practice of this method, which will be considered in detail in the next chapter, requires several different types of mentality. To gather evidence and verify conclusions, there must be keen observers, ingenious experimenters, and painstaking classifiers. To frame concepts and explanations, there must be imaginative theorists and hair-splitting logicians. To put the findings of the others to everyday use, there must be down-to-earth pragmatists. Whether for applied or for purely research purposes, everyone who employs the scientific method falls under the heading of scientist: the laboratory technician and the collector of zoo specimens as well as the Einstein and the Darwin; the social scientist, as well as the physicist, chemist and astronomer.

The strength of six million

In this broad sense there are, all told, some six million scientists in the world today. Only a few hundred thousand hold Ph.D.s or teach at universities. Only a few hundred, in or out of the academic sphere, are recognized as prime creative talents. But science is such that it makes full use of all its echelons. Without an army of intellectual foot soldiers to man the oscilloscopes and test tubes, the great laboratories and research institutes would quickly close their doors.

So many and varied are the grades and shades of scientist that an attempt to catch all their likenesses in a single portrait would be futile. Accordingly, in the pages that follow, research scientists will be given center stage—particularly those who practice in the old established fields where the scientific method has been pursued furthest. By general agreement, these are the members of the scientific community who set the standards for all the rest, and who come closest to the image of the scientist entertained by the public at large.

The popular view of the scientist has changed considerably since World War II. In 1943 General Leslie R. Groves, welcoming his staff of nuclear wizards to Los Alamos, opened his remarks with a joke that probably reflected the true opinion of most practical men of affairs at the time. "At great expense," he said, "we have gathered here the largest collection of crackpots ever seen." By 1958 a sampling of adults across the country showed that this traditional prejudice had begun to disappear. Only 40 per cent of those polled endorsed the statement that scientists "are apt to be odd and peculiar." Fewer still—25 per cent—felt that "scientists always seem to be prying into things they really ought to

DISTRIBUTION OF SCIENTISTS throughout the continental United States is highly uneven, as the map below shows. The nation has about half a million scientists, but the two most populous states, California and New York, each have 12 per cent of the total, while sparsely populated Arkansas has only 0.2 per cent. Five states (the blacks and grays) command the services of 40 per cent of U.S. scientists.

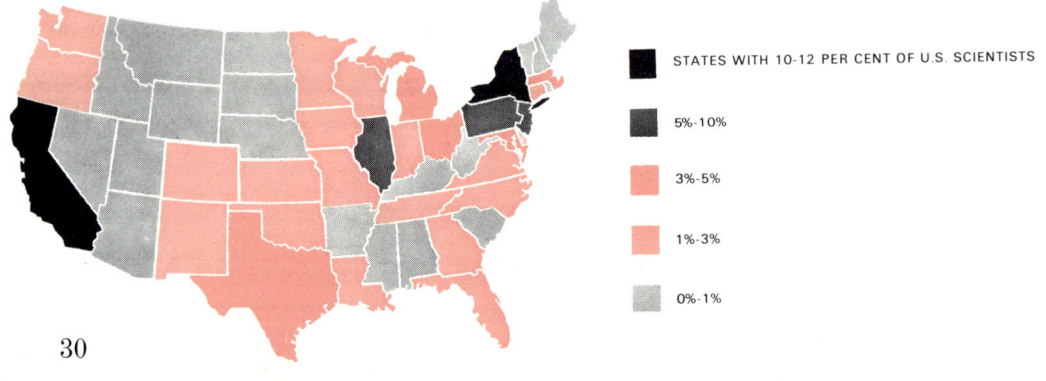

stay out of." A resounding 88 per cent believed that "most scientists want to work on things that will make life better for the average person."

If anything, the image of the scientist has been burnished almost too brightly in recent years. A survey of 35,000 high school students, conducted by the anthropologists Margaret Mead and Rhoda Métraux in 1957, indicated that the younger generation regards the scientist as a veritable saint. They imagine him as elderly or middle-aged, sometimes stooped and tired, but always courageous, dedicated, stubbornly optimistic, intelligent, meticulous and patient. Typical of their comments were the following: "The future rests on his shoulders." "Through his work people will be healthier and live longer . . . our country will be protected from enemies abroad."

Perhaps understandably, the youngsters revealed an altogether different reaction when asked to picture themselves in the role of this prim paragon. They now saw his work as dull, lonely drudgery, underpaid and possibly dangerous. They felt that he might lose his religious faith. They feared that he might be accused of treason. They believed that at the very least he would neglect his family or bore them stiff.

Since not enough students are going into science to meet future needs, educators alarmed by such attitudes have begun to campaign for a more realistic knowledge of the scientific life and personality. As White House assistant McGeorge Bundy wryly noted before the American Association for the Advancement of Science in 1962: "Scientists are people, a fact which is frequently forgotten but verifiable experimentally."

Rigors in a vineyard

The misconception of the scientist's lot as a tedious one is the easiest to scotch. To acquire his education he must indeed study hard, and to pursue it professionally, especially in basic research, he is likely to continue working hard. But if these be rigors, they are self-imposed. To an extent enviable by workers in other vineyards, he is passionately interested in what he is doing. The best research scientists operate at a level of intensity and excitement seldom matched in other callings. They love their art and, what is more, they are paid to pursue it, usually with complete freedom from direction. Study grants, lecture invitations, teaching appointments, travel funds and international conferences keep them on the move and in the professional swim.

In the freedom of the scientific world individuality thrives and stereotypes wilt. Some scientists serve humanity self-effacingly; others are flamboyant egotists. Some are convivial souls who like to rub shoulders and tell jokes; others play the misanthrope, wishing well only of minerals, vegetables and the more nearly extinct species of animals. One scientist may scoff at religion, while another prays daily for guidance.

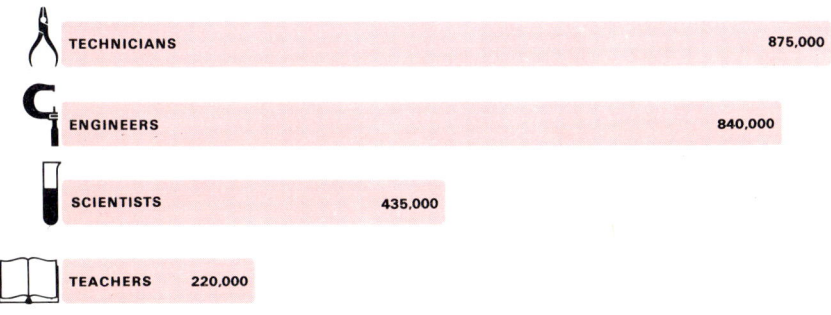

THE SCIENTIFIC COMMUNITY consists of armies of technicians and engineers, a division of scientists and a regiment of science teachers in high schools. The bar graph at left shows the relative sizes of all four groups in 1960. Although their absolute numbers change from year to year (1960's total of 2.37 million is unofficially estimated to have grown to 2.7 million by 1964 and is projected to four million by 1970), their relative proportions are expected to remain the same.

One may read nothing but technical periodicals and the Sunday comics, while another quotes Dante or Shakespeare like an Oxford don. One may wear loafers and substitute talcum powder for socks, while another is always turned out as impeccably as a banker.

In pursuit of a pattern

Notwithstanding the many outward differences between men of science, psychologists have discovered by extensive testing and analysis that there is a certain distinctive combination of traits which characterizes the scientific personality. One of the most searching studies of this personality pattern was conducted over a three-year period in the early 1950s by the clinical psychologist Anne Roe, wife of the paleontologist George Gaylord Simpson. Dr. Roe's subjects were selected by university committees all over the country as topflight research scientists. They were a diverse lot: anthropologists, botanists, geneticists, paleontologists, psychologists, chemists, biochemists, physicists and astronomers —64 of them in all, averaging 48 years old. Dr. Roe put each one through a battery of psychological and intelligence tests. She then double-checked her findings by a less detailed testing of 382 other scientists.

The most striking characteristic shared by all of Dr. Roe's subjects was a formidably high level of intelligence—an average I.Q. well up in the 150s as compared to about 100 for the ordinary citizen. She gave them three separate tests: spatial perception, mathematical aptitude and verbal aptitude. They did extremely well at spatial perception, brilliantly at mathematics, and best of all in the verbal test—thereby belying the notion that men of science are generally inept with words.

In areas more subjective than intelligence, Dr. Roe was able to reach a number of tentative conclusions through lengthy personal interviews and such psychological standbys as the Rorschach inkblot and Thematic Apperception tests. Since generalizations about types of personality are often suspect, it is worthy of note that her findings have been confirmed again and again by other investigators working with wholly different groups of subjects. Many of these subsequent projects were encouraged by the Government's National Science Foundation. Hoping to improve its selection of fellowship applicants, the NSF sought some means of recognizing those with the greatest potential as creative scientists. More than 50 research reports on "The Identification of Creative Scientific Talent" were presented in the late 1950s at three NSF conferences in Utah. From Dr. Roe's conclusions, and those revealed at the Utah meetings, a coherent picture of the U.S. research scientist emerges.

• He has a "general need for independence, for autonomy, for personal mastery of the environment." He resists pressures to conform in his thinking. He is attracted by facts or ideas which seem mutually contra-

If $x + 3y = 7x + 5y$, then $x/y = ?$
(A) −3 (B) −1/3 (C) −1/9 (D) 1 (E) 3

APTITUDE TESTS, like the math problem above (with five answers to choose from) and the two puzzles at right (two figures in each row are identical), were especially devised for 64 scientists from four widely varying fields, to determine the mental aptitudes of various kinds of scientists. Physicists in the group were excused from the math part of the test, on the grounds that any math designed for the others would be too easy for them, but even without them, the rest of the group scored very superior in mathematical ability. The group as a whole was superior in perception of spatial relationships and very superior in verbal skills. Correct answers to the three problems are given on page 34.

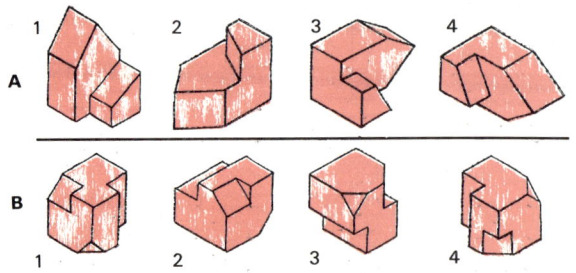

dictory, and finds it challenging to reconcile them. He likes to pit his gifts against uncertain circumstances in which he feels that his own efforts may be a decisive factor.

• In his attitude toward his work he is neither self-sacrificing nor particularly unselfish, but simply happy and intent. Vacations often strike him as annoying interruptions. Passive recreations, such as moviegoing, are likely to bore him.

• Emotionally, he is at the same time stable and sensitive. Though often inclined to be highly critical of others, he avoids personal controversies. When hurt socially he is likely to move away from the offender and repress any outward show of hostility. But instead of brooding, he tends to "switch off" the outside world by submerging himself in thoughts about science. The social scientist, however, is an exception to this rule; he is sociable almost by definition. It may be evidence of an occupational hazard that he also has a relatively high divorce rate—in Dr. Roe's sampling, 41 per cent as opposed to 15 per cent for the biologist and only 5 per cent for the preoccupied physicist.

• The research scientist has a strong ego. This sometimes makes him overly dignified and likely to keep an unduly tight rein on himself. He is not usually impulsive or talkative. If he is a biologist, rather than a physical or social scientist, he is apt to be especially cautious about weighing words and avoiding conjectures.

• In his younger years the research scientist usually develops a precocious self-confidence about solving intellectual problems. His family, almost invariably, sets great store by book learning.

• In adulthood, he tends to be open-minded about religion. He is not notably agnostic or atheistic. He is inclined, however, to be somewhat careless about church-going and skeptical about the details of any one doctrinal orthodoxy.

• As a member of the community, he comes off with high marks. In Dr. Roe's words, scientists "pay at least as much attention to civic duties as the average man does; they do not enrich themselves at others' expense; they and their families rarely become public charges, and the more violent crimes are practically unknown among them."

• Over-all, the scientist's personality is remarkably similar to the profile that psychologists draw of creative artists or writers. He is slightly less sensitive, and slightly less burdened with feelings of guilt and anxiety. But he shares with other creative groups a strong tendency toward introversion and self-sufficiency.

The psychologists' portrait of the research scientist has been augmented by statistics. In general, Protestants and Jews produce a larger quota of scientists than the Catholics do. The reason for this imbalance appears to be a matter of historical and cultural tradition. The scientist, in his emphasis on individual inquiry and on the constant re-examination of accepted truth, would seem to be a cultural child of the Reformation.

Until World War II a disproportionate number of American scientists came from relatively poor families or professional families; from small liberal-arts colleges, state universities and high schools rather than Ivy League colleges and prep schools; from farms and small towns rather than big cities; from the West and Middle West rather than from the Northeast. In the last two decades, as the aura of science has generally brightened, these regional and class differences have begun to disappear.

Increasingly, the scientist has become more than a distinctive personality, more than a fascinating public image. He has become a whole new force in the progress of civilization; an unprecedented source of cultural change; a dispenser of more miraculous inventions and more difficult ideas than anyone, scientists included, can keep up with and fully understand.

Quartermaster for mankind

He derives his public prestige and influence largely from the fact that technology breeds on basic research. Our affluent society, our high productivity, our time-saving machines all hinge on his discoveries. He has become the quartermaster for mankind, issuing jets to generals, detergents to housewives, automation to manufacturers and new channels of communication to politicians.

As a class, therefore, research scientists find themselves playing a role unusual for people of their creative temperament. They have become a powerful elite—but one that is altogether different from other groups of oligarchs in previous eras. They have no vested interest in the status quo; indeed, they cannot help but change it by the very practice of their craft. Nor is wealth, social position or self-interest of much help in advancing their cause. The talisman of their power is talent, and a steady stream of it must flow into their brain-power pool lest it stagnate.

As a result, few scientists ever indulge the luxury of being clannish and exclusive. In a constant quest for Fermis and Einsteins, they advertise the delights of their calling widely. They keep no trade secrets and welcome all apprentices. Astonishingly often in history one scientist will acclaim the ideas and help in the training of a younger man destined to unseat and overshadow him. The tradition is a venerable one. When Isaac Newton was a 27-year-old graduate student at Cambridge, his chief mentor, Isaac Barrow, then only 39 himself, voluntarily resigned in Newton's favor from the Lucasian chair of mathematics.

National boundaries and ideological barriers have almost as little meaning for the scientist as fine lines between social classes. Scientific genius crops up without regard for country, language or color. The laws

TEST ANSWERS to the problems on page 32 appear below. With the physicists unscored, biologists did best on the math section. Psychologists were a close second, with anthropologists trailing badly. Physicists did best on the perception of spatial relationships, with psychologists hot on their heels, biologists next and anthropologists last. Theoretical physicists and anthropologists ranked highest in verbal skills.

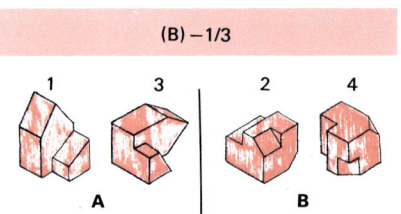

of nature operate identically in all lands. They are everywhere for the finding—accessible to investigators of every nation.

The very vocabulary of the scientist is international. His equations are universally applicable; the technical words he uses, being derived from Latin or Greek, are at least pan-European. There are fundamental reasons for this preference of his for a scientific Esperanto. In grappling with stars, geologic time, the instincts of animals or the impalpable particles of the subatomic underworld, he must try always to think in terms of realities that transcend local landscapes and languages. No wonder, then, that he sees himself as part of an international—or even interplanetary—community.

Before World War II, when scientific accomplishments still had relatively little direct or sudden impact on the fate of nations, scientists took their internationalism for granted. At an international scientific conference in those less complicated times, they could assume that they were being told all the known truth and nothing but the known truth about the subject at hand. Now, at least in certain areas, it is necessary to interpret what is said in the light of security regulations and diplomatic maneuvering. Natural laws are still universal, but they sometimes have to be discovered twice, once in one camp and once in the other. Though scientists have become far more politically sophisticated than they once were, most still look upon ideological obstacles to international discourse as necessary evils.

The price of cosmopolitanism

Scientists have suffered considerably for their cosmopolitan convictions, and the general public has learned slowly to be tolerant of them. In the U.S., during the tribulations of atomic physicists before Congressional investigators in the 1950s, the entire scientific community sometimes felt itself under popular suspicion. Harsher trials beset scientists in totalitarian lands. During the Stalin regime, some Soviet scientists were jailed because of their contacts in the West, while others were ordered to renounce foreign scientific theories and invent patriotic Marxist dogmas in their stead.

When persecuted by dictators for ideological reasons, scientists defend themselves articulately, but if no one listens, they ultimately retreat—often to a nation's detriment. Nazi Germany lost some of its finest scientific brains by emigration, and Allied weapons research gained accordingly. In the U.S.S.R. during periods of repression, controversial Russian scientists have often turned their attention to gadgeteering or inoffensive theorizing. Thus, at a time when atomic engineering and highly abstruse solid-state physics throve, the forbidden in-between ground of relativistic quantum mechanics sank out of sight. In the life

ANTOINE LAVOISIER, the 18th Century chemist who discovered the nature of combustion *(left, below)*, was a victim of the French Revolution—one of the many scientists throughout history who have been involved in the politics and conflicting ideologies of their times. An engraving of the period shows soldiers of the Revolution *(right)* bursting into Lavoisier's laboratory while he gestures to friends and assistants nearby. Lavoisier was guillotined for having served the monarchist government.

sciences, applied research in genetics dried up and geneticists retreated into theoretical contemplation of the fruit fly.

As they have become ever more aware of the crucial value of scientific discovery, rulers both autocratic and democratic have come to realize that they tamper with the scientist at their own risk. The liberalization of climate which took place in the Soviet Union under Khrushchev was in part due to scientists' pressure. In the West physicists exercise more and more influence over military and industrial decisions. Biologists sit in at conclaves concerning health, welfare, parks and wildlife. Sociologists, psychologists and economists participate in judicial and budgetary proceedings. However much he may prefer his old contemplative ways, the scientist is now deeply involved in the world of affairs.

Never has society lionized a more creative, or a more unpredictable, breed of hero. Intellectually, the scientist is likely to have more in common with the philosopher or the scholar than with the traditional man of action. But the overwhelming triumph of technology has drawn him increasingly into board meetings and councils of state. It is a reasonable assumption that in the decades ahead he will come to wield more power than any elite class has ever done in the past—more than the feudal lords of the Middle Ages, the merchant princes of the Renaissance or the tycoons of the Industrial Revolution.

Instruments: Tools for Scientific Conquest

Man has devised a wide assortment of scientific instruments in the course of his investigations of the universe. Some, like the astrolabe *(opposite)* used by ancient astronomers, have helped with the basic scientific task of measuring the visible world. Two of the most important—the microscope and the telescope—have extended the visible world by thousands of times. Still others enable scientists to explore such impalpables as the atmosphere and such invisible forces as electricity. Renaissance craftsmen set high standards of beauty and accuracy for the making of instruments, and instrument-makers have contributed importantly to the growth of science ever since. As science became more complex, it moved its growing paraphernalia into laboratories—where both ideas and instruments could be pooled. By the early 20th Century, in such centers as the Cavendish Laboratory at Cambridge University, the first atomic-research tools were forged.

"MATHEMATICAL JEWELS"
The medieval astrolabes pictured on the opposite page were so useful to astronomers that they were known as "mathematical jewels." This instrument, going back to the Second Century A.D., was mounted on one side with a revolving star map and crossbars to take readings from the rim. The other side *(lower right)* had a sighting bar, and helped determine latitude and time of day.

NOBLE ASTRONOMER
Tycho Brahe (1546-1601), a member of the Danish nobility, plotted the positions of 777 stars with his new and accurate instruments. His nose was cut off in a youthful duel, and he wore a false one of gold and silver—discreetly left unnoticeable in this contemporary portrait.

HISTORIC SEXTANT
Tycho had several sextants, probably including this one *(below)* made in Augsburg of wood and brass. After his death, it is believed to have gone to his young assistant, Johannes Kepler.

Precise Tools for the Naked Eye

Early astronomers evolved many instruments besides the astrolabe for charting accurately what could be seen in the heavens with the naked eye. "Naked-eye astronomy" reached a high point of precision with the work of a 16th Century Danish astronomer, Tycho Brahe *(above)*.

Tycho was only 14 when an eclipse of the sun fired his interest in astronomy and he began his lifelong work of finding and correcting errors in current astronomical tables. He was the first, for example, to allow for the effects of refraction by the atmosphere in his calculations of star positions. He recognized, however, that some earlier errors were caused by the small, imprecisely calibrated instruments of the time. He became convinced that larger, more carefully designed instruments were needed.

As a young nobleman who had already acquired a brilliant reputation in astronomy, Tycho was brought to the attention of King Frederick II of Denmark, who backed his work with royal funds. Tycho went to Augsburg and Nuremberg, where the finest craftsmen were to be found, and there supervised the construction of a number of outsized astronomical instruments. Some, like the sextant pictured at right, were inventions of his own; others, like the equatorial armillary and azimuth quadrant on the opposite page, were enlarged and improved versions of devices that astronomers had used for centuries. Made to Tycho's meticulous specifications, they set new standards of accuracy in astronomical research.

In an extravagant use of King Frederick's money, Tycho built this elaborate observatory in 1576, on Hveen Island off Copenhagen.

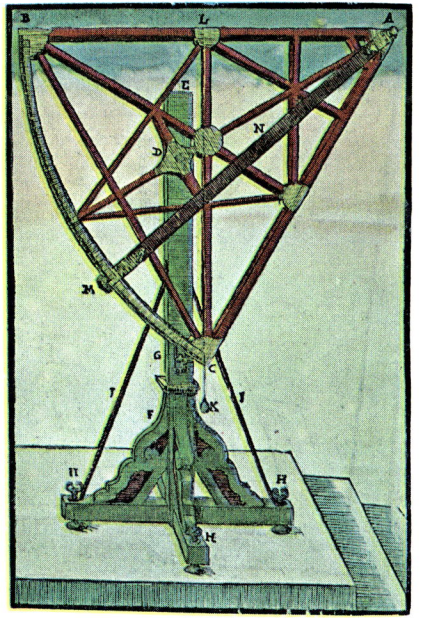

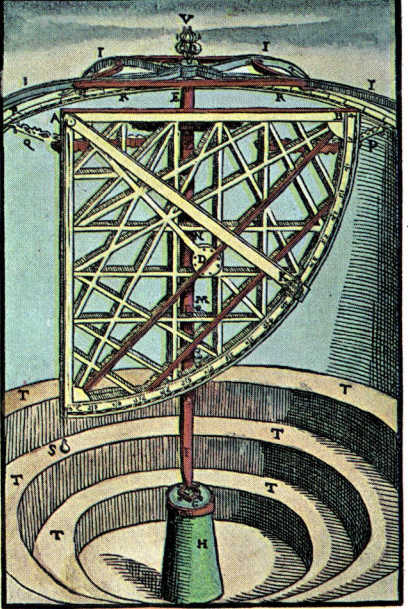

TOOLS FIT FOR A SPLENDID OBSERVATORY

Tycho wrote a book about his instruments, illustrated with drawings of them, including the three above. The sextant pictured at left measured about six feet along one side (AB). The huge equatorial armillary *(center)* was about 16 feet high and had its own enclosure in Brahe's observatory. Observers mounted the tiered steps at the base to sight into the heavens along the plane of the great revolving ring. The picture at right is of a 10-foot azimuth quadrant, accurate to 1/240 of a degree. Visiting scientists are reported to have jumped for joy on seeing Brahe's instruments.

NEWTON'S TELESCOPE
Newton's reflecting telescope *(above)* is about six inches long. Light enters, at right, is reflected and focused by a mirror at the far end and bounced by another to the eyepiece at the side.

Tools to Extend Man's Vision

Dutch spectaclemakers are generally credited with inventing both the microscope and the telescope within the span of a few decades—the microscope in 1590, the telescope in 1608. Men had been stargazing for centuries, and the telescope was put to almost instant use. Galileo got wind of the device in Italy, and by 1609 had built his own to discover a new celestial world. The first telescopes were of the refracting type, which used a lens to focus an image. Early lenses caused distortion, however, and in 1668 Isaac Newton devised a reflecting telescope *(left)*, in which a mirror does the focusing. Many modern telescopes are of this type.

It took longer for microscopes to be put to serious use. They were called "flea glasses" because they were used to look at insects, but in 1665 the English scientist Robert Hooke published *Micrographia (below)*—the first documentation of the microscopic world, including metals and plant life.

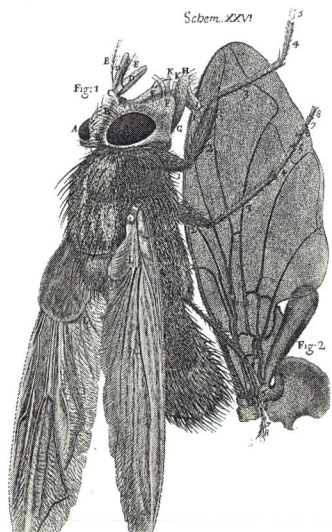

DOCUMENTING THE MICROCOSM
Among the items Hooke observed with his microscope and drew in detail for his book were the common blue fly *(above)* and thin slivers of cork *(below)*. He described the tiny cavities in cork as "little boxes or cells," and the basic units of life have been called cells ever since.

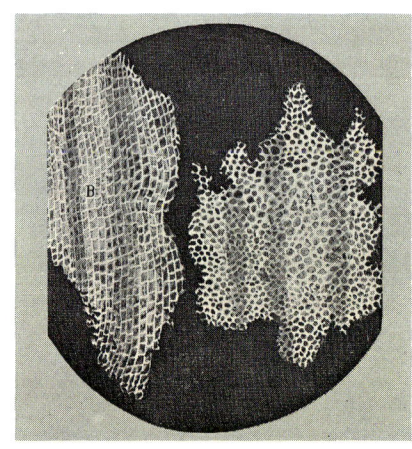

HOOKE'S MICROSCOPE
Robert Hooke designed this microscope and used one similar to it in preparing his book *Micrographia (above)*. Two lenses were placed at top and bottom of the six-inch tube, mounted on a movable ring. The object being viewed was stuck on a pin in the base, and the instrument could be lengthened with drawtubes. Holly and a razor's edge were objects studied by Hooke.

THE PROLIFERATION OF A NEW TOOL
By the end of the 17th Century, microscopes were no longer a rarity. The simple one on a lacquered wooden box at left was made about 1690; the rest are 18th Century products. The unadorned instrument at upper right is one of many like it produced in Nuremberg about 1760, but in an age when all instruments were handmade, many were still individually designed and richly decorated for their purchasers. The tube of the center microscope is covered with Moroccan leather; the one at right is gilded bronze; below it is an instrument of rosewood and ivory.

The Art of Instrument-making

During the Renaissance, instruments were custom-made for the wealthy few, and money and craftsmanship usually combined to make them beautiful. Elaborate portable sundials—precursors of the pocket watch—incorporated their own compasses and astrolabes, so that the instruments could be oriented to tell time at different places on the earth.

The new class of instrument-makers prospered and their shops became centers of scientific shoptalk. In the 18th Century machines were first used in the making of scientific tools. These machine-made instruments—cheaper and more accurate than their handmade counterparts—filled the needs of the growing numbers of amateurs and dilettantes who followed the scientific fad *(below)*.

PRINCELY SUNDIALS

Almost every Renaissance prince had his *cabinet de mathématiques*, full of handsome gadgets like these *(above)*. The "astronomical compendium," or pocket timepiece *(top, center)*, of about 1600, has a compass on top, a ring sundial and a plumb bob. The 16th Century brass box *(center)* has a compass on its top and several sundials for use in different latitudes.

INSTRUMENT-MAKER TO THE PUBLIC
Jesse Ramsden, shown above with a machine he invented for inscribing scales on metal, was one of the new 18th Century craftsmen who made instruments in mass instead of on order. His machine replaced hand-engraved lines with those spaced with mathematical precision.

POKING FUN AT A FAD
English gentry look through the wrong ends of telescopes and examine an armillary sphere with a magnifying glass in this copy of an 18th Century copper engraving *(below)*. Books were one means by which interest in science spread, but cheaper instruments hastened the process.

KEPLER'S CLOCKWORK GLOBE
This exquisite gold-plated globe *(above)*, made in Augsburg in 1586, is said to have been used by Johannes Kepler. It was almost a miniature observatory: the large celestial sphere showed the stars, and was topped by a small armillary sphere. The smaller terrestrial globe underneath is mounted over a compass. Clockworks turned the large globe and chimed on the quarter hour.

SCIENCE AND SENTIMENT
This theatrical scene, painted in 1768 by English artist Joseph Wright, was called *A Philosopher Shewing an Experiment on an Air Pump.* The same experiment was first performed by Boyle with a pump based on Guericke's. As air from the glass globe was evacuated, the lark was seen to pant, go into convulsions and die —while a tenderhearted girl covers her eyes.

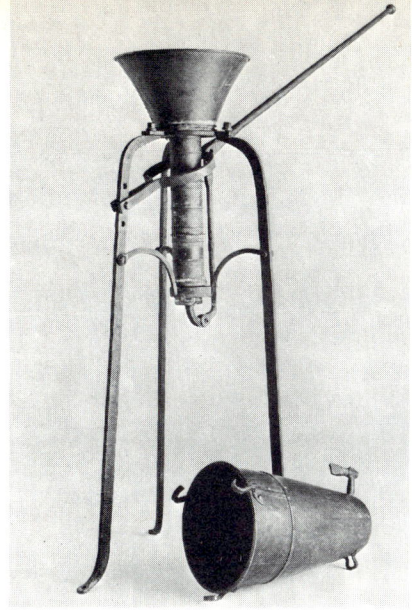

WATERTIGHT AIR PUMP
In this improved air pump of Guericke's, the empty container was filled with water and hung below the cylinder. The cylinder was thereby submersed, and air leaks were prevented.

Investigating Air and Electricity

The two devices illustrated on these pages—a static-electricity generator and an air pump—were among the first tools that enabled scientists to experiment with unseen phenomena such as air, vacuums and electricity.

The air pump, which could suck air from a container, was invented in the mid-17th Century by the German burgomaster Otto von Guericke. With it, Guericke demonstrated that air had weight, and roughly calculated its density. One of his most celebrated experiments was with the "Magdeburg Hemispheres"—two hollow bronze hemispheres perfectly fitted together. He pumped the air from them, and two opposing teams of eight horses failed to pull them apart.

Guericke also built the first static-electricity generator—a ball of sulphur as big as a baby's head, mounted on an axle. After being charged by rotation against the hand, it attracted light objects such as feathers, produced sparks and proved an electric charge could travel. The machine roused little scientific curiosity. Over half a century went by before scientists took much interest in electricity, and built complex generators like that below to produce greater quantities.

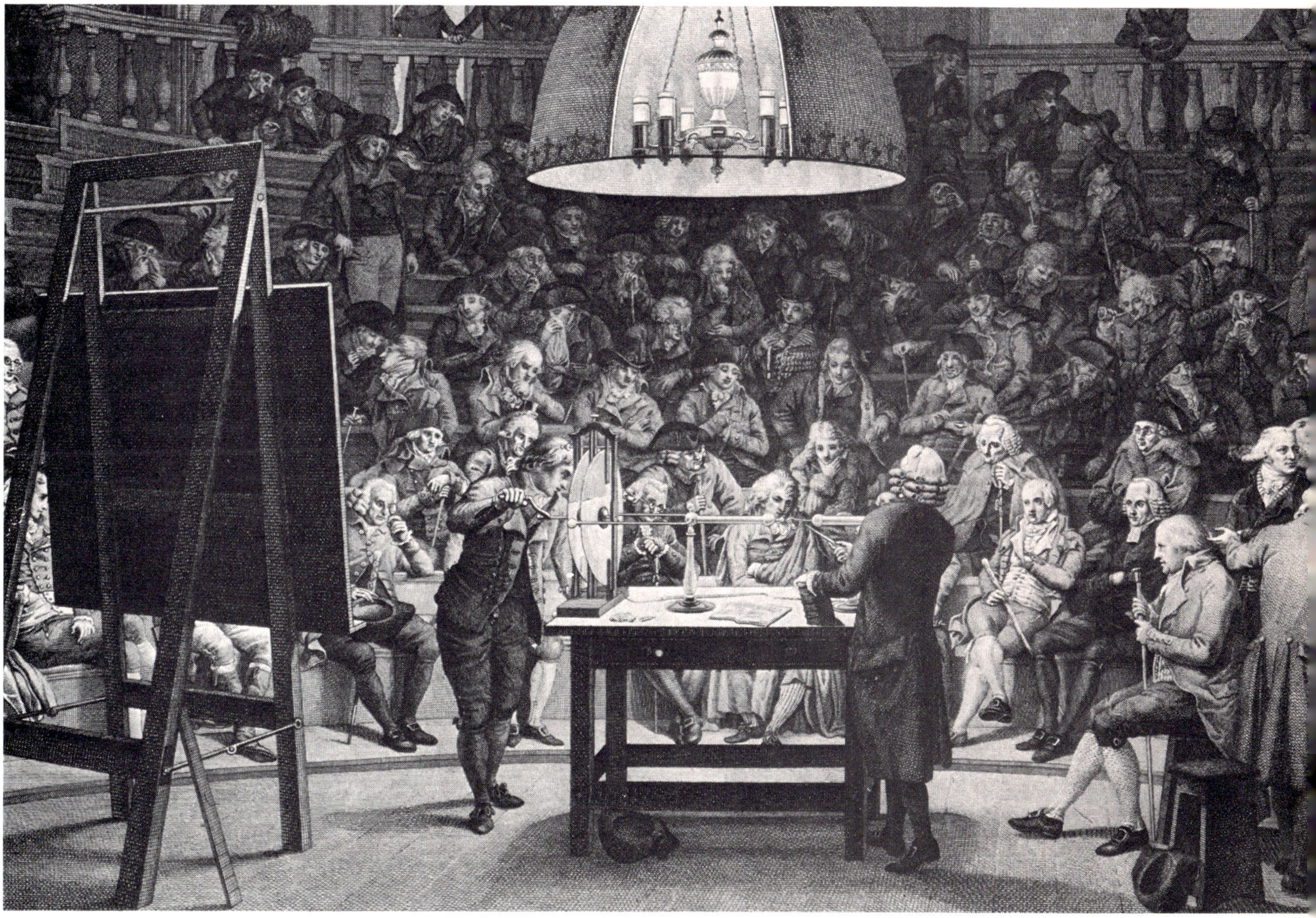

ELECTRICITY MADE VISIBLE
This 18th Century engraving shows a young physics professor demonstrating an "electrical machine" before a private science society in Amsterdam. The glass plates at the left end of the device are revolving against leather rubbing pads and generating static electricity. The electricity could be made to produce sparks, or be stored in collecting devices called Leyden jars.

Putting Chemistry into the Lab

The origins of the modern chemistry laboratory go back to the workrooms of medieval alchemists, who devised skilled techniques and serviceable tools in the effort to turn base metals into gold. As interest in the physical world grew in the 16th and 17th Centuries, the alchemists' tools and techniques were turned more and more to the investigation of the fundamental properties of matter.

As interest in the new field grew, particularly in Germany, chemists began working together to share facilities as well as ideas. In 1825 the first large modern lab was established by Baron Justus von Liebig, professor at the University of Giessen. Liebig's lab *(above)* was for nearly 30 years the mecca of the world's chemists, who came to learn—and carried his methods back to their own countries. There the study of organic chemistry took shape. A host of chemical compounds of commercial value was discovered, and the basis of Germany's coal-tar and dye industries was laid. Students were drilled in qualitative and quantitative analysis, prepared compounds and carried out original research—a program followed by teaching laboratories ever since.

A PROTOTYPE WORKROOM
The drawing above, sketched in 1842, shows the main room of Liebig's chemistry laboratory at Giessen—where professors and students alike wore hats to keep the ashes from charcoal burners out of their hair. The arrangement of the lab and its equipment soon became standard for chemistry teaching labs: workbenches with cupboards, shelves, water supply and sink; at the back, air shafts for dealing with noxious gases. On the tables are such items as beakers, flasks, evaporation basins, bell jars and mortars, all still familiar to chemistry students today.

ALCHEMY'S LEGACY
The instruments at right are reproductions of those Liebig himself used. In the center is a distilling apparatus. Flanking it are two retorts and some tongs. All of them bear a strong resemblance to the equipment pictured in medieval illustrations of alchemists' worktables.

Nursery of the Atomic Age

Most of the great moments of early atomic physics occurred in the Cavendish Laboratory, founded in 1871 at Cambridge University. Under a succession of brilliant directors, it became traditional at the Cavendish lab for a young physicist to make his own instruments—reportedly sometimes out of string and sealing wax. In this informal atmosphere of ideas and homemade hardware, the electron, the atomic nucleus and the neutron were discovered. In 1911 the Wilson cloud chamber was invented, making the tracks of atomic particles visible, and in 1932 one of the young physicists who had helped put it together pulled the switch to start the first large atom smasher *(opposite)*.

ELECTRON HUNTER
Appointed director of the Cavendish lab at age 28, J. J. Thomson *(above)* ran it on a budget smaller than its present phone bill. He is shown here with the machine with which he discovered the electron in 1897. His discovery was the first indication that the atom was not an indivisible entity, and gained him a Nobel Prize —one of 14 awarded to Cavendish scientists.

BAILIWICK OF GENIUS
In this cluttered corner at the Cavendish Laboratory, Ernest Rutherford discovered the atomic nucleus and also explained the disintegration of radioactive elements. Rutherford, who succeeded Thomson, was director from 1919 to 1937, perhaps the lab's most illustrious period. He is remembered chiefly as a physicist, though he won a Nobel Prize for chemistry in 1903.

MAKESHIFT ATOM SMASHER
John Cockcroft and Ernest Walton assembled this primitive particle accelerator, described as being made of "plasticine, biscuit tins and . . . sugar crates." The wooden crate was a darkroom, in which the experimenter crouched behind a curtain to view the tracks of particles.

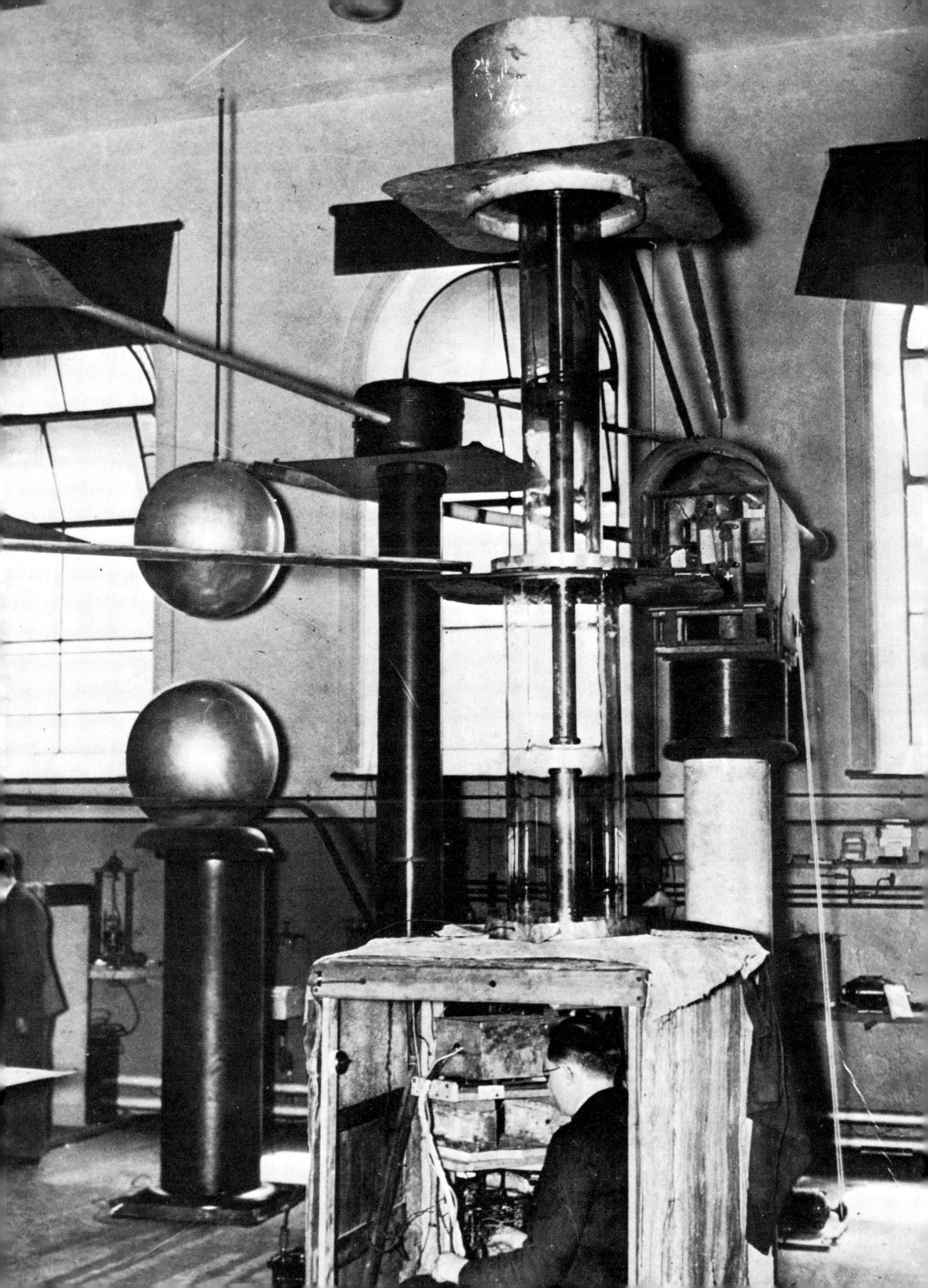

3
The Scientific Method

WE HAVE ONLY TO TAKE A WALK on a spring morning, or watch the waves break on a shore, or gaze at the stars over the back porch, to be reminded that the world around us is a tapestry intricate beyond compare. But it is one thing simply to admire the façade, and quite another to delve behind it. This is the goal which the scientist perennially seeks, and which sets him apart from his fellowman.

The fabric of outward appearances is no easy one to penetrate. It is apparently seamless, and its very beauty is distracting. Where, and how, does the scientist begin to see through it? How can he glimpse, beyond it, the law and order of the universe? What gives him his unique confidence? The answers lie in the scientist's particular style of inquiry, one which distinguishes him from all other kinds of thinker. This approach, which he has developed over the ages in his quest for knowledge, has come to be known as the scientific method.

Few aspects of science are more intrinsically elusive, more difficult for the layman to grasp, yet none better repays the effort—for the scientific method goes to the very heart of all scientific activity. In this chapter an attempt will be made to convey something of its vast scope, by describing its virtues and limitations, by analyzing its various steps, and by discussing its broad implications.

To begin with, it should be noted that the term "scientific method" is itself somewhat of a misnomer. It is not a method in the sense of a formal procedure. It furnishes no detailed map for exploring the unknown, no surefire prescription for discovery. It is, rather, an attitude and a philosophy, providing guidance by which dependable over-all concepts can be extracted from impressions that swarm in on man's senses from the outside world.

So all-encompassing is the method that it can be employed fruitfully by scientists of every specialty and every type of talent. Its practitioner may be the kind who seems to be forever turning over rocks to expose what lies beneath, smashing things open to count and catalogue their interior contents—always obsessed with facts. Or he may be, in the tradition of a Newton or an Einstein, someone we see as a dreamer, pursuing visions, spinning gossamers of equations high in an abstract sky—working not with facts but with ideas that are seemingly born of sheer creative fancy.

It is the great triumph of the scientific method that it enables these two extremes of talent, the data-gatherers and theory-makers, to complement each other. It bridges the chasm between facts and ideas, between meaningless helter-skelter and satisfying order, between everyday applications of science and the theories from which they are born. It gives the scientist the discipline to distinguish between ideas which are relevant and useful, and those which are empty and misleading. It enables him to

A CONTINUING QUEST
Dr. Irving S. Wright, a leader in cardiovascular research, checks blood samples for the effects of anticoagulant drugs on the clotting that causes strokes. Testing and retesting are integral features of the scientific method. Wright's studies, launched more than two decades ago, demonstrate that thus far only one patient in three or four responds to anticoagulants.

exploit those fugitive moments of intuition and insight which are as indispensable to science as to any other mental endeavor.

With its virtues, the method has certain natural limitations. It cannot replace the inspiration of Archimedes discovering a basic law of hydrostatics while sitting in his bath. It cannot conjure up the good luck of Alexander Fleming chancing on penicillin. It cannot hasten the slow process of intellectual growth and seasoning. In short, it cannot create science automatically any more than the theory of harmony can write a symphony or a Navy manual can win a sea battle.

A guide for the inquiring

The scientific method is the offspring of a branch of philosophy called epistemology (from the Greek *episteme*, "knowledge," and *logos*, "theory"). From Plato on down, men have pondered the question of what knowledge consists of, whence it comes and how we acquire it. In the 17th Century the English epistemologist Francis Bacon, focusing on scientific knowledge in particular, attempted to provide a tidy blueprint for its acquisition. He offered the scientist a fourfold rule of work: observe, measure, explain and then verify. By the 19th Century a more sophisticated version of the method was being propounded: pose a question about nature; collect pertinent evidence; form an explanatory hypothesis; deduce its implications; test them experimentally; and then accept, reject or modify the hypothesis accordingly.

But observe what? Measure what? How frame the hypotheses or explanations? Because there are obviously no pat answers to these questions, present-day epistemologists doubt that creative scientists can be hedged about by hard-and-fast instructions. One man may pick up a butterfly and notice the peculiarities of its wing markings. Another may wonder about the ratio of its flying surface to its body weight. The first may go on to enunciate a new theory of insect camouflage and the second a new theory of insect aerodynamics, but neither gains his insight by measuring all available evidence indiscriminately or by testing every arbitrary hypothesis that flits into his mind. One may arrive at the answer he seeks after a good night's sleep. The other may badger his inspiration for many years before he sees the light. Or then again, on a lazy day, both may simply appreciate the beauty of the butterfly and let it flutter off.

History may wait decades for an encounter of the right man and the right butterfly on the right day. No single investigator is likely to proceed from an initial observation to a finished theory about any significant subject in the course of his lifetime. When Newton drew up his fundamental laws of motion and gravitation in the 1680s, he capped an investigation that stretched far into the past: back through Galileo's experiments with falling cannon shot and Kepler's formulation of laws for the movements of the planets; back through the shrewd speculations about force and relativity made by a group of medieval thinkers known as the Impetus School; back through the deductions of ancient Greeks about geometry and space; back through the stargazing of the Babylonians; all the way back to Stone Age observations on the motion of projectiles. When Darwin arrived at his concept of evolution, he answered questions about changing forms of life which had been posed by scientific thinkers as long ago as Aristotle. Astronomers today still consult the readings of star positions made by their ancient colleagues in Samarkand and Sumer.

Any program of creative research in science moves ahead through tentative, chancy, inspired steps taken by many men over many years. A theory dimly glimpsed in the 19th Century may emerge into full view only after it has been framed in mathematical patterns of logic worked out in the 20th Century. An experiment recognized as vital may endure lengthy postponement while the apparatus is being devised to perform it. The physiologist Wilton Earle, who pioneered in growing live human tissues in bottles outside the body, was once held up for an entire year for lack of a flask stopper which would not poison his delicate sterile cultures. It took him months to ascertain that the stoppers he needed should be made of silicon, and months more to persuade a manufacturer to turn them out.

These obstacles which impede research, as well as the strokes of luck and intuition which help it on its way, are basic to all human enterprise. What is unique to scientific endeavor is the linkage it forges from ideas to facts, and from facts to ideas.

The fruits of conjecture

Nothing is more astonishing about science than its ability to make imaginative conjectures and then convert them into tangible realities which no one had previously suspected. Out of Maxwell's equations of electromagnetism came radio and television. Out of Einstein's formulas on matter and energy came the atomic bomb. When the scientific imagination works in the opposite direction, to crystallize theories out of facts, the transformation is equally dazzling. Here, in one decade, lies a musty lot of old bones and odd birds, and there, in the next decade, stands a theory of evolution. Here is a pile of dirty pitchblende and an endless labor of refining to do, and there is radium and an awareness of the atomic nucleus.

To understand the remarkable interaction that science sets up between observable facts and abstract ideas, it is necessary to consider these questions: What constitutes a fact of science? What constitutes an acceptable theory? What happens intellectually in the process of bridging the gap between the two?

The facts with which science works are called data, from the Latin meaning those things which are given or granted to us—granted through our observation of the world around us and our absorption of its unceasing flow of sights, sounds, smells, tastes and tactile sensations. It is this kind of outer experience that is employed in the making of science, rather than the kind of inner experience generated out of man's hungers, fears, loves and other urges. Insofar as he can, the scientist, when considering evidence, excludes emotion and bias from his thinking. True, various kinds of excitement may help to fire his thoughts. But he tries to discount his sense of righteousness or of sin; his simple pains or joys in living; his repugnance for ugliness or attraction to beauty. That certain something which makes the word "home" convey more than "house" falls outside his present professional competence. As takeoff points for his inquiry, the scientist accepts only those ingredients of experience which bear discernibly on objective knowledge—knowledge as distinct from hope, desire, contemplative enjoyment or mystical rapture.

Restrictions of a realm

This restricted realm of experience in which the scientist operates is described by epistemologists as cognitive (from the Latin *cognoscere*, "to find out," "to know"). It admits only those perceptions which we receive directly through our senses, plus the purely logical processes by which we select, judge and reason. It contains such items as the chest of drawers which we have seen and touched and therefore *know* to be there, even when we have turned our backs on it; the gallon of gasoline which we have tested and therefore *know* will run a car for a certain amount of mileage; the sun which we *know* will rise again tomorrow because we have seen it rise so many times in the past.

The raw stuff of science lies strewn about on the surface of experience like so many nuggets on a sandbar. Scientists sometimes call the field from which it is taken the "protocol" plane. The word stems from a practice of Greek authors in classical times. When a writer planned a work, he would first paste on the manuscript a sheet of papyrus listing the topics which he expected to explain in the text. This preliminary sheet was known as the *protokollon*, from *protos*, "first," and *kolla*, "glue." Hence it seems appropriate to refer to the raw sensory materials of knowledge as the protocols, or forewords, of science.

When he selects his protocol facts, the scientist tries to define them as precisely as he can through measurement. The kinds of measure he employs vary a great deal. The astronomer classifies a star by its brightness and spectral color. The chemist discriminates between substances

MEASUREMENT OF TIME has become a finer and finer process through the centuries. An early way of telling time was by observing the position of the sun in the sky. Sundials, appearing about 1500 B.C., made possible a more exact division of the day into hours. Galileo's discovery of the constant timespan of the pendulum's swing led to the development of far more accurate timepieces in the 17th Century than had ever existed before. Ammonia clocks, first devised in 1948, tell time by vibrations within the ammonia molecule. When a radio wave of the correct frequency stimulates the molecule, a single nitrogen atom *(pink dots)* vibrates back and forth through a plane created by three hydrogen atoms *(white dots)*. Its vibrations are so regular variations may amount to only one second in 3,000 years.

ANCIENT ASSYRIAN OBSERVING THE SUN

SUNDIAL

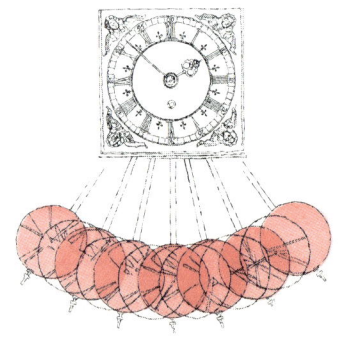

PENDULUM CLOCK

by their boiling points, acidity and so forth. The physicist defines a push or a pull in terms of force which can be registered on a spring balance.

By such means a scientist can assign a numerical value to a protocol fact which heretofore had only sensory or esthetic value. He changes it from a qualitative fact, which is subjective and private, into a quantitative fact, which is objective and public—in other words, communicable. Sensations of hotness and dampness, for example, can be converted into universally understood quantities like 90° Fahrenheit or 70 per cent relative humidity.

Some scientists believe that certain protocol facts may never be defined completely by measurement, but will always require description in qualitative terms, verbal rather than numerical, ambiguous rather than precise. Many others, however, feel that the purely verbal definitions still remaining in science are temporary expedients. In physics, terms like force, temperature and energy could be defined only verbally in the early days before 1600; now they are entirely quantitative—specific symbols in equations. By the same token, in biology, the fact that traits are passed on from parent to child was at first recognized in a single broad word, "heredity." Then Gregor Mendel analyzed it more precisely, and broke it down into hereditary units called genes, which can be observed and measured by their effects, and so reckoned with mathematically.

Through the use of standards and gauges—through measurement, in short—the scientist distills from his private sensations an essence which all men can inspect, verify and use. The Stone Age primitive who put his finger in boiling water may have voiced his equivalent of "hot!" When he gazed at a new moon he may have made a mental note of the passage of time. But these were his perceptions alone. Today, having invented thermometers and scales of temperature, we have a mutual understanding of how hot "hot" really is. Having measured time with clocks, we have a mutual appreciation of time which stands up under scientific comparisons.

Grist for the theorist's mill

In science, every sensation which is expressed in terms of an instrument of measurement is thereby converted into a concept. Such a concept is said to be "operationally defined," a phrase coined by the late Professor Percy Bridgman of Harvard. Once measured, protocol facts provide grist for the mill of the theorist. Examining various numbers arrived at by measurement, he can seek out relationships between them, which he expresses in formulas. Readings of temperature and pressure made on the air in a tire, for example, can be converted into the basic law of gases: that pressure rises or falls proportionately with tempera-

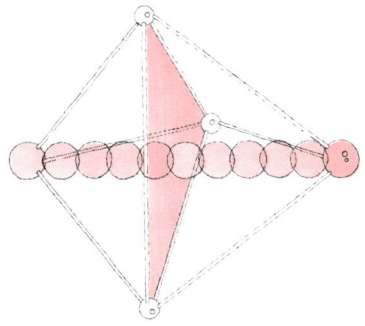

AMMONIA MOLECULE

ture. Time, having once been regulated by the tick of a clock, becomes a mathematical concept independent of personal time as gauged by the heartbeat. As such it can be bred with other measurements—combined with miles or kilowatts, for instance, to yield miles per hour or kilowatt hours.

Readings on gauges do not automatically produce laws or basic understanding. They may lead nowhere—unless the scientist sees a pattern in them. And such patterns are difficult to discern unless measured values are expressed in suitable units. Many of our present units, our weights and measures, are legacies from the early years of scientific inquiry, and do not suggest any truly fundamental ideas. The metric system is based not on natural units meaningful throughout the cosmos but on 18th Century measurements of the circumference of the earth. The Fahrenheit temperature scale, based on a zero point defined by a haphazard mixture of salt and ice, does not suggest the true nature of temperature, which we now know to be molecular agitation. If scientists were to use a thoroughly modern set of standards, they might insist on the Kelvin scale of temperature calibrated from the point of $-273°$ C., the so-called "absolute zero" at which molecules cease to move, and merely quiver; they might organize a system of time tuned to the rate of expansion of the universe.

The fundamental numbers in nature

Through increasingly refined measurements, the scientist seeks and finds ever more fundamental "numbers" in nature. It is by this process that he has discovered such crucial quantities as absolute zero, the speed of light in a vacuum, the mass of the hydrogen atom. Picking out fundamental numbers from a welter of measurements, coming to have a familiar feel for them, and eventually evolving from them some concept about nature, constitute the most difficult and creative part of a scientist's work. All too easily, he may pursue blind alleys or go not quite far enough in his analysis. Before Newton, for instance, Galileo, Kepler and William Gilbert, the great Elizabethan student of magnetism, all came close to grasping the essentials of gravitation, but all veered off through various misconceptions concerning what physicists now call force.

In entering the domain of concepts, the scientist mentally shifts gears, as it were. Whereas in dealing with protocol facts and measurements he has used inductive reasoning, he now applies deductive reasoning. In reasoning inductively, he has worked his way logically from particulars to generalities. "All the swans I have ever seen are white; therefore all swans are white." Or: "I have never seen a fire-breathing reptile; therefore dragons do not exist." But these generalizations that he makes are never entirely certain. After all, swans have been found in Australia which are black, and someone might conceivably turn up a reptile which had learned the sideshow trick of swallowing fire.

After enough tests, however, if a generalization seems unexceptionable, the scientific theorist proceeds to treat it as true—at least tentatively. He takes it as a premise and reasons out possible consequences that might arise from it. If his premise is true, the conclusions he derives from it will be true too. In applying this type of reasoning—deductive logic—he can draw on the symbols and patterns devised by mathematicians, the purest scientists of all. When Einstein envisaged a new theory of space and time, he was able to take over a mathematical structure previously conceived by a school of 19th Century non-Euclidean geometers. When modern biochemists theorize about DNA molecules, they shape their information with tools already hammered out by probability theorists and topologists.

The deductive stage of a scientific investigation is sometimes called the "C-domain," the "C" representing concepts. In this domain the scientist constructs theories more or less at will and explores their implications. Each new twist he gives to his hypotheses puts him into a new orbit of possible consequences. Each fresh go-round broadens his perspective on the protocol facts from which he started. After enough orbits, he may begin to discern new factual possibilities not apparent to him in his original observations or measurements.

A crucible of questions

As he formulates his final theory, the scientist subjects it to intensive criticism. Seeking to make it as useful as possible, he asks himself: Is this proposed law universal throughout the extent of space and the passage of time? Does it lead anywhere? Does it predict one state of affairs as arising out of another? Can it be transposed from one frame of reference to another and still remain valid? And finally, because of his innate passion for orderliness, his esthetic appreciation of things which are meet and fitting, he asks: Is this theory as elegant as possible? Could I formulate it more succinctly?

Now comes the moment of verification and truth: testing the theory back against protocol experience to establish its validity. If it is not a trivial theory, it suggests the existence of unknown facts which can be verified by further experiment. An expedition may go to Africa to watch an eclipse and find out if starlight really does bend relativistically as it passes the edge of the sun. After a Maxwell and his theory of electromagnetism come a Hertz looking for radio waves and a Marconi building a radio set. If the theoretical predictions do not jibe with observable facts, then the theorist has to forget his disappointment and start

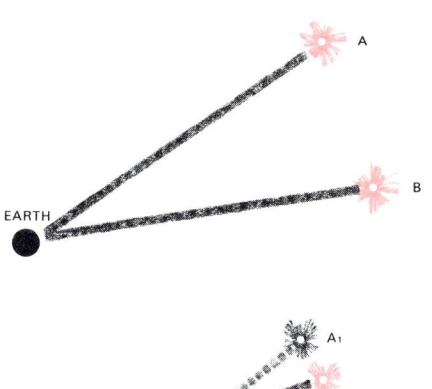
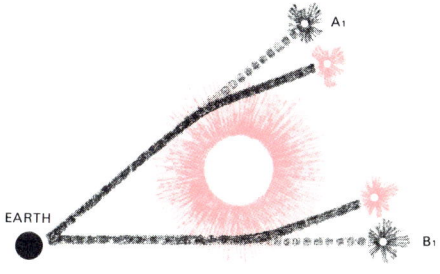

PROOF of Einstein's theory that light bends when it passes through a gravitational field was made possible by a total eclipse of the sun in 1919. The diagram at top shows the normal paths of light from stars A and B to earth. The sun passed between these paths during the eclipse, and in the darkness, the two stars were photographed. True to Einstein's prediction, their rays were bent *(above)*. The deflected rays *(solid lines)* caused the photographic images to appear as though they came from points A_1 and B_1 along the projected dotted lines. The passing of the sun, in other words, made the stars appear to move.

all over again. This is the stern discipline which keeps science sound and rigorously honest.

If a theory survives all tests and is accepted into the canon of scientific law, it becomes a fact in its own right and a foundation for higher spires of thought. Abstract though it may be, a proven theory can suggest new hi-fi sets or hybrid cattle just as surely as do experiments with electricity or stockbreeding. It serves as a starting point for new theories just as surely as any experience on the plane of protocols. Galileo's formula for the increasing speed at which a body falls freely near the surface of the earth became a single example of Newton's law of gravitation. Newton's law, in turn, became a single special case in Einstein's theory that gravitation is a manifestation of the geometry of space and time. At this moment some child in a hamlet in Maine may be preparing himself for the work of constructing a "unified field theory" of both atom and cosmos, in which Einstein's sweeping concepts of relativity will appear as mere details.

In full bloom, rooted in protocols, stemmed with measurements and headed with theories, a science becomes a complete organism. As is suggested by the marginal diagram at left, every part of it communicates with every other part wherever located—on the P-ground of protocol fact, in the M-air of measurement, or higher still in the C-space of concepts. Vital juices flow from experimental instruments to mathematical symbols and back again. Past ideas remain as germinal in present science as do the cells of childhood in the body of an adult.

The dilemma of success

The astonishingly effective tie-in that the scientific method creates between literal-mindedness and imaginative symbolism has produced problems by its own luxuriant success. Today, the roots of the tree of science delve so deep into the subatomic underworld, and the branches reach so high into the realm of the galaxies, that scientists are not always sure what it is that they are dealing with. Is an evanescent atomic particle, artificially created and artificially detected, a protocol fact or a concept? Is the expansion of the universe, deduced from spectra of faint heavenly bodies, a reality of nature or a description imposed on nature by the mind? Are the human senses and intelligence, developed for a middle-sized world of tables and chairs, capable of exploring an atom many billions of times smaller than man, or a starry galaxy hundreds of millions of trillions of times larger? Is it possible that, in the words of the British geneticist J.B.S. Haldane, the universe "may not only be queerer than we imagine, but queerer than we can imagine"?

Obviously, after their recent successes in both atomic and cosmic matters, scientists are not about to give up as hopeless their epic quest for

THE SCIENTIFIC METHOD can be diagramed to show the two-way relationship between "protocols"—or the raw data of sensory experience—and scientific theory. The scientist selects certain protocols, for example sounds, and measures them. This act converts them into quantitative concepts, in this case sounds of specific decibels. When these concepts acquire logical and mathematical relationships, they form a theory of sound. Once formed, the theory must hold true for all auditory sensations on the protocol plane.

knowledge. Nonetheless, there has been a marked increase in scientific modesty during recent decades. Many scientists of the 19th Century believed that the number of scientific facts to be learned was finite. They often felt that they would someday achieve absolute truth and ultimate understanding. Today, however, their successors speak only of reaching "partial understanding," of continually approaching truth but never grasping it completely.

An old debate renewed

The difficulties which now beset the scientific imagination have revived an issue which Aristotle and Plato were arguing about 2,000 years ago. What is reality? Which is more real and dependable: an idea generated by the mind, or an impression of something outside, generated through the senses? Rationalist philosophers like Plato and Descartes have inclined to the view that ideas exist independently. Empiricist philosophers like Aristotle, Bacon and Hume have felt that ideas do not exist except as they are suggested by observable facts. Taking a position somewhere between the two schools, the great 18th Century German metaphysician Immanuel Kant once mused: "Concepts without factual content are empty; sense data without concepts are blind. . . . The understanding cannot see. The senses cannot think. By their union only can knowledge be produced."

Under the aegis of the scientific method, the rationalists and empiricists of science—the theorists and the experimentalists—have got on so successfully together that, for the most part, they have left philosophy to the philosophers. But they have come to acknowledge that the age-old debate about the nature of reality is fundamental to science today. Ideas do, indeed, lead a life of their own which is independent of observation and experiment. The concept of a circle, for instance, might still have its own validity even if the universe were devoid of any circular shapes. The ratio of the circumference to the diameter of a circle, designated as π by the ancient Greeks, and known as pi to all schoolchildren since, has been calculated to 100,000 decimal places, yet no one, to this day, has ever seen a circle perfect enough to exemplify this ratio beyond about six decimal places. The mathematical concept of a circle, in short, is more dependable and certain than any circle which can be drawn with a compass or observed in nature.

That the ideas of science are created by the mind, as well as discovered in the outside world, was brought home dramatically to 20th Century scientists when they realized that they had been attributing to subatomic particles—such as the electron—a number of qualities which have meaning in our everyday world but little or no meaning in the microcosm of the atom.

Scientists think of the electron as the carrier, within the atom, of negative electrical charge. The rest of us, when we think of it at all, usually envisage streams of pellets in power lines. The electron is too small to touch, taste, smell, hear or see, even with a microscope, but we nonetheless imagine it to be essentially concrete and palpable. After all, electrons can be shot from the nozzle end of a particle accelerator as if they were so many bullets; the fragments of their targets can be traced in a photographic emulsion almost as vividly as if they were splinters from a doorframe shattered by machine-gun fire.

But do electrons really have anything in common with pellets or bullets? Imagine what would happen if we could shrink a little round red ball and make it as small as we wished. When its diameter got down to one hundredth of an inch, the ball would cease to be visible to the naked eye. It would still have color, roundness and size, however, as could be determined through a magnifying glass. When the magnifying glass no longer sufficed, a microscope would still serve as a legitimate extension of human vision to follow the ball's shrinkage. But no matter how superbly effective the instrument of observation, the ball would begin to lose its color when its diameter shrank to less than 25 millionths of an inch; since this is about the wavelength of red light, anything smaller cannot reflect light as we know it. The ball would become not black, but simply unimaginable in terms of color.

The end effect of shrinkage

As it continued to shrink, the ball would remain detectable by shorter waves, which can be received by instruments other than the eye. Ultraviolet rays would continue to bounce against it until it was less than one millionth of an inch in diameter. Eventually the ball would lose any appreciable shape, size or position. At 40 trillionths of an inch in diameter—the size of an electron—it could no longer withstand the impact of the most lightweight ray or particle which could be hurled in to rebound from it. Any probe, devised experimentally or imagined theoretically, would succeed, not in "seeing" the ball, but in knocking it galley-west.

This is precisely the position in which the electron itself is found—that is, in no position at all. The very act of finding an electron is synonymous with dislocating it. It is impossible to know, at one and the same time, where an electron was when you found it and where it is when you become aware of the finding. It is, in sum, impossible to pinpoint the place and speed of movement of any one electron at any one instant.

The scientist attributes to the electron many valid descriptive qualities—spin, charge, mass—which are deduced rather than observed directly. If he says, however, that the electron has no position or definite size, or that it behaves sometimes like a particle, sometimes like a wave, he is using words which refer to sensory qualities meaningful in our own world but altogether irrelevant in the world of the microcosm. To look for color in an electron is like trying to smell a sunbeam.

The dilemma posed to the human senses by the individual electron—or by any other subatomic particle—was met head on in 1927 by the quantum physicist Werner Heisenberg, when he enunciated the so-called "uncertainty principle." According to this principle, we can never justifiably attribute to the electron, at any one time, all the everyday descriptive qualities of commonsense objects. The very act of finding it traveling at a given speed will throw its whereabouts into question. The very act of determining its whereabouts will give it a speed or direction that is unknown. Locating electrons in space and time can be achieved only by applying the laws of probability. As with human beings, the electron population is statistically predictable, but the behavior of capricious individuals within the group is not.

A tradition gone awry

The concept that chance governs the behavior of electrons has played hob with traditional notions that a particular cause produces a particular effect. In the subatomic world, one event is not a *necessary* consequence of another event, but merely a *possible* consequence—with a calculable probability.

Many scientists feel that Heisenberg's principle of uncertainty is intrinsic to the whole universe; that the intimate details of the cosmos have a fuzziness which science will never be able to clarify or overcome. Other scientists, perhaps a minority, feel that the uncertainty may be the result of human limitations rather than of any unpredictability on the part of nature itself. In effect, Heisenberg's dictum may simply be a reminder, to the human being who seeks to capture the spirit of the cosmos, that attempts to visualize its minutiae are futile. The Biblical injunction against the making of graven images may still hold.

More and more, scientists regard their concepts as "models" which are analogies of nature rather than photographic likenesses of it. Furthermore, different models may be used to represent the same concept. The atom, for instance, has sometimes been pictured as a solid sphere armed with hooks; at other times it has been thought of as a pudding of positive electric charge filled with negatively charged raisins—the electrons. In the model of the atom which has proven most widely useful—the one conceived by Niels Bohr in 1913—it is compared to a miniature solar system. If we were to magnify such a solar system 10 trillion times, we would get a wild picture of the sort of wild picture which scientists must contend with. The "sun" in this system—the nucleus of the atom—is as big as a marble, but unimaginably denser. The nearest "planet"—elec-

A "PROBABILITY CHART" like the one above is one way physicists chart the elusive location of an electron at a given moment. Each white dot in this diagram represents a chance that the electron is in the vicinity. As the concentration of white dots shows, there are many more chances that it is near the center, or nucleus, of the atom, although there is some chance that it has ranged to the outer edge.

tron—orbits a third of a mile away. It is lighter, fluffier and larger than the "sun"—about the relative size of a 50-foot gas balloon. If an earthling were placed beside the same model and magnified the same number of times, he would be 10 billion miles tall, and his head would be far beyond the outermost planets.

The incredible complexities of the electron, though they may stagger the imagination of the layman, have proved immensely instructive to the scientist. He continues to look for manifestations of nature. But when he finds them he knows that they are simply manifestations— facets of reality which happen to suit a particularly useful, and often beautiful, idea which his mind has created. His method, in short, is a means of creating new ideas and then of selecting the ones which best fit the appearances of nature.

In a manner of speaking, the scientist stands in a house of mirrors —mirrors which he himself has built, and which he himself keeps improving so that they give increasingly detailed and accurate reflections. Reality continues to elude him, but he can see it darkly in one or another of the flawed glass surfaces before him. Some of the mirrors give one distortion, some another; all offer only reflections. By piecing together these reflections, the modern scientist seeks ever better approximations of truth, but he no longer expects to see truth naked.

The Pursuit of Omega Minus

A good example of the scientist's search for laws of natural order has been the effort to make some sense of the swarm of entities that came tumbling out of the atom's nucleus when the physicists turned on their high energy accelerators in the 1950s. Previously the atom had been believed to be a neat structure with a few simple parts; now scores of subatomic bits streamed from it. The presence of these bits were so baffling that physicists labeled the situation a "crisis in physics." At the same time, physicists were sure that some order could be made of this enigmatic overabundance, and that the order, once found, might lead to vital secrets about the origins and nature of the universe. Here is the story of the search for Omega Minus, a nuclear particle whose existence was predicted by a theory that established some order among the welter of particles. The theory, itself difficult to grasp, paradoxically makes the universe easier to understand.

PORTRAIT OF A PARTICLE

The slanting white line third from the top in the picture on the opposite page is one of the most significant photographic images in all science. It is the bubble-chamber track of the Omega Minus particle, actually a string of bubbles, enlarged about 500 times. The snowy effect between the lines is caused by the enlargement. An actual-size reproduction appears on pages 70-71.

MURRAY GELL-MANN
Gell-Mann, pictured above and at Geneva below, was 31 when he published his eightfold-way theory at Caltech in January 1961. He has contributed ideas to particle physics since age 21.

The Call to the Hunt

At a 1962 gathering of the world's nuclear physicists in Geneva, a 32-year-old Caltech physicist named Murray Gell-Mann proposed a search for a nuclear particle he called Omega Minus. Its existence was predicted, he argued, by a theory formulated independently by himself and another physicist, a 37-year-old former Israeli Army engineer named Yuval Ne'eman. This theory Gell-Mann called "the eightfold way." Gell-Mann had derived the poetic name from the eight quantum numbers in nuclear physics (described below) and from Buddha's eight ancient prescriptions for the good life.

When Gell-Mann and Ne'eman proposed their eightfold way, physicists already suspected that atomic particles were somehow related to each other—that perhaps they had evolved from a common origin, just as animals evolved from some common primordial ancestor. But before Darwin could conceive his biological theory of evolution, earlier men first had to classify animals into species according to resemblances and differences. A similar classification was needed for nuclear particles, and this was offered by the eightfold way.

YUVAL NE'EMAN
By a coincidence typical of science, Ne'eman published his theory—similar to Gell-Mann's—in the same month in *Nuclear Physics*. He too was present at the Geneva meeting *(below)*.

FROM AN OLD TO A NEW ATOMIC ORDER

THE CLOCKWORK OF THE ATOM
The eightfold way deals in particles, forces and "quantum numbers." Particles are the separate identifiable bits into which the atom can be broken. They include the electrons, protons and neutrons, long thought to be the atom's only constituents; the photons and mesons, discovered later, that were found to play a role in "glueing" the atom together; and, finally, the short-lived entities—over 150 have been identified—that are found after an atom is smashed by an accelerator.

An electromagnetic force exists between any two objects that are electrically charged—pulling positive and negative charges together, causing like charges to fly apart. A more powerful so-called "strong force" holds the nucleus together in spite of electromagnetic repulsion.

Quantum numbers are used to rate the properties of the various particles. These properties include such things as mass, spin and electric charge as well as others not clearly understood.

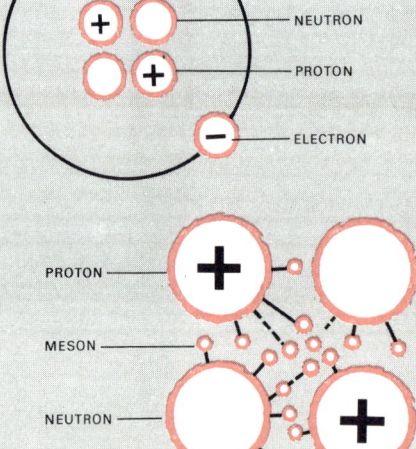

A NEAT AND SIMPLE ATOM
By the 1930s, physicists knew that the atom was made up of a nucleus containing positively charged protons and uncharged neutrons, circled by negatively charged electrons *(top, left)*. The electrons were held in their orbits by electromagnetic force. Ordinary light was another manifestation of electromagnetism, in the form of waves instead of a steady field of attraction or repulsion. By this time evidence had accumulated that light came in separate "bundles" called photons, or light quanta. Physicists reasoned, and later demonstrated, that photons must also be "carrying" the electromagnetic force within the atom, perhaps moving between protons and electrons like a kind of magnetic pendulum, pulling them together before they could fly apart.

Eventually, these protons were found to behave as though they were individual particles.

The nature of the mysterious "strong force" or "nuclear glue," remained to be explained.

NICHOLAS SAMIOS
A 30-year-old experimental physicist from Brookhaven National Laboratory on Long Island, Samios heard Gell-Mann's talk, was immediately interested in hunting for Omega Minus.

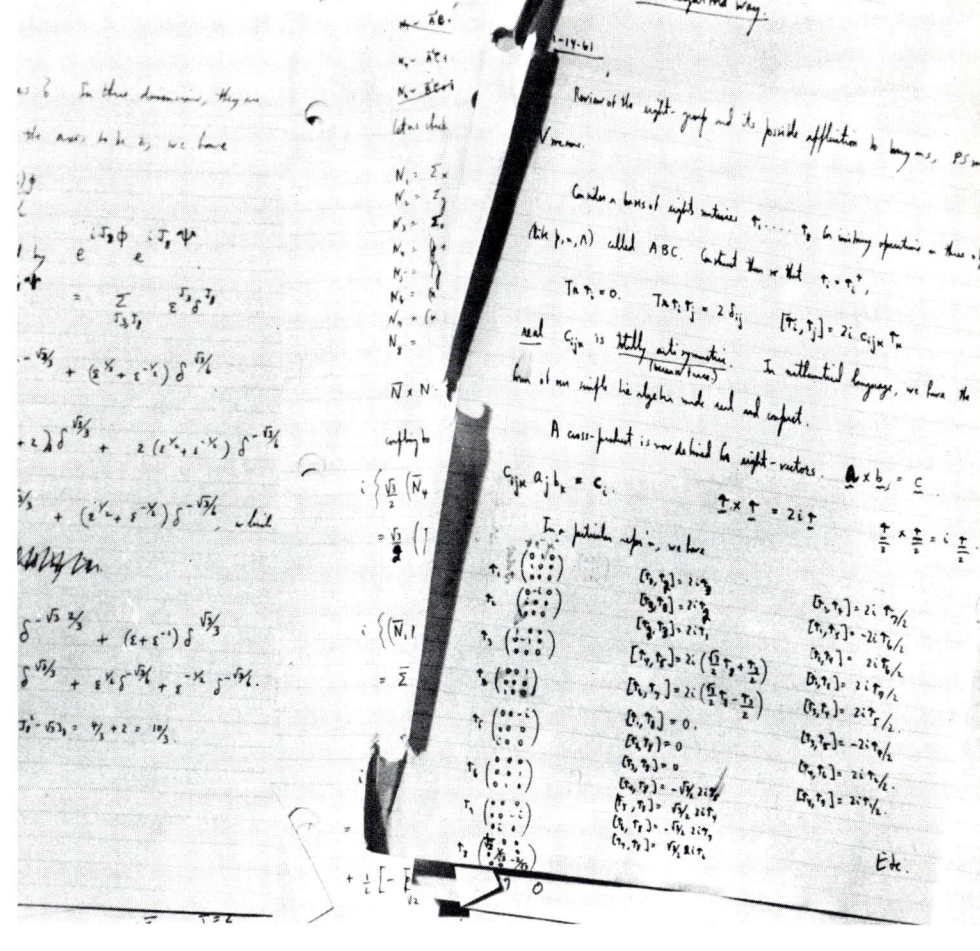

OLD MATH FOR A NEW THEORY
Above are two pages of the notebook, titled *The eightfold way*, in which Gell-Mann worked out his idea in mathematics. He had to rediscover an obscure mathematical system invented in the 19th Century in order to manipulate numbers in groups of eight, since each interacting nuclear particle had eight quantum numbers. Independently, Ne'eman did the same.

RDER FROM THE EIGHTFOLD WAY

sing the photon as a model, Japanese physist Hideki Yukawa in 1935 proposed that her particle "carried" the strong force beveen the particles in the nucleus *(bottom, left)*. ventually, Yukawa's particles, called mesons, ere discovered in not one, but three varieties.

In the '40s and '50s physicists began blasting a host of other particles out of atomic nuei. Many of them showed enough similarities o that they could be classified into families, or multiplets," whose members differed from ch other only in electric charge. These multiets were given Greek letter names: Delta, gma, Psi, Kappa, and so on.

Gell-Mann and Ne'eman theorized that many these particles have additional properties, milar in some ways to electric charge, that ve rise to the strong force, just as electric arge itself gives rise to the electromagnetic rce. However, instead of being limited to two operties only—positive charge and negative charge—as is the case with electromagnetic force, eight properties were necessary to account for the strong force's peculiarities. These new properties were given quantum number ratings. The eightfold way went on to establish a theory to predict what happens when two particles collide—whether the interactions among the eight properties would create strong forces that would make the particles repel each other or attract each other to form a new particle. Furthermore, just as electric charge allowed particles to be grouped into "multiplets," the eight quantum numbers made it possible (through complex calculations) to group particles into "supermultiplets." The average mass of each multiplet *(horizontal rows at left)* is about 147 units heavier than the average mass in the multiplet below. Accordingly, Gell-Mann and Ne'eman were able to predict that still another missing particle with a large mass—the Omega Minus—would top off the pyramid.

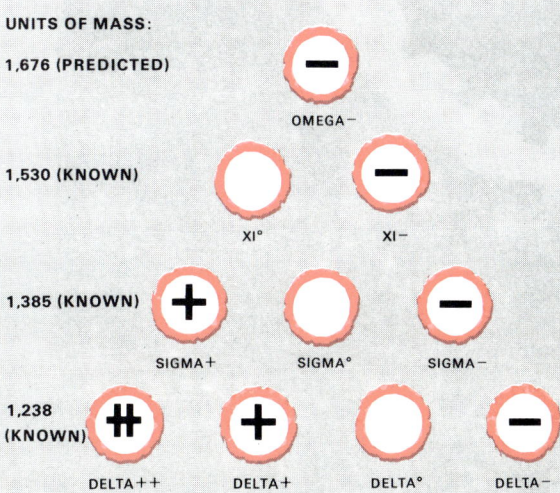

UNITS OF MASS:

1,676 (PREDICTED) — OMEGA−

1,530 (KNOWN) — XI°, XI−

1,385 (KNOWN) — SIGMA+, SIGMA°, SIGMA−

1,238 (KNOWN) — DELTA++, DELTA+, DELTA°, DELTA−

THE CHAMBER IN THE MAKING
Only a mammoth magnet was in place, ready to be fitted with Brookhaven's new bubble chamber, when Nicholas Samios returned from Geneva in 1962 with the proposal to search for Omega Minus. The experiment had to wait until late 1963 while the chamber was completed.

Large Weapons for a Small Quarry

The most powerful machines of experimental physics were needed to prove the existence of Omega Minus. One was in operation in 1962 at Long Island's Brookhaven National Laboratory: the alternating gradient synchrotron or AGS *(diagram below),* an atom smasher half a mile in circumference. The second, to be used in tandem with the AGS, was also being forged at Brookhaven: an 80-inch hydrogen bubble chamber *(opposite),* still under construction, whose relatively great length would be needed to reveal an Omega Minus particle's creation.

A bubble chamber is a piece of apparatus that almost puts experimenters in visual touch with the infinitely small particles ripping through their equipment at near the speed of light. Huge magnets surrounding the bubble chamber change the track they take. The paths the particles follow, their collisions and the particles that emerge from these collisions are traced by tiny hydrogen bubbles that form in their wakes—bubbles that can be photographed. The photographs tell the physicists what particles they are dealing with.

After the new Brookhaven bubble chamber was finished, a group of physicists, headed by Ralph P. Shutt, hunted for Omega Minus, with Samios in direct charge of the experiment.

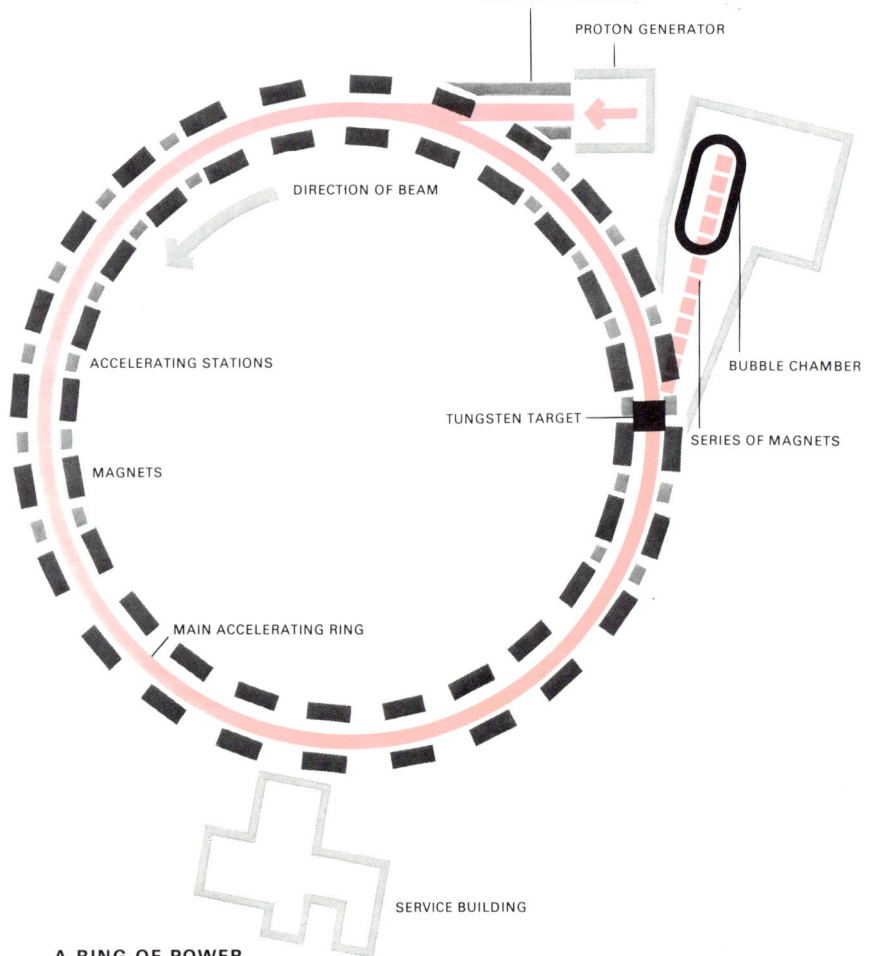

A RING OF POWER
In Brookhaven's AGS, protons from an atomic nucleus are used as bullets. First a generator separates the protons, then a linear accelerator kicks them into the main accelerator ring, where magnets confine them and accelerating stations jolt them to near the speed of light. They smash into a piece of tungsten and fragments of tungsten atoms are sent toward the bubble chamber.

FILTER FOR FRAGMENTS
The particles shattered from the tungsten are sent through holes in several magnets *(above).* The magnets weed out the particles, sending on only the ones wanted for the experiment.

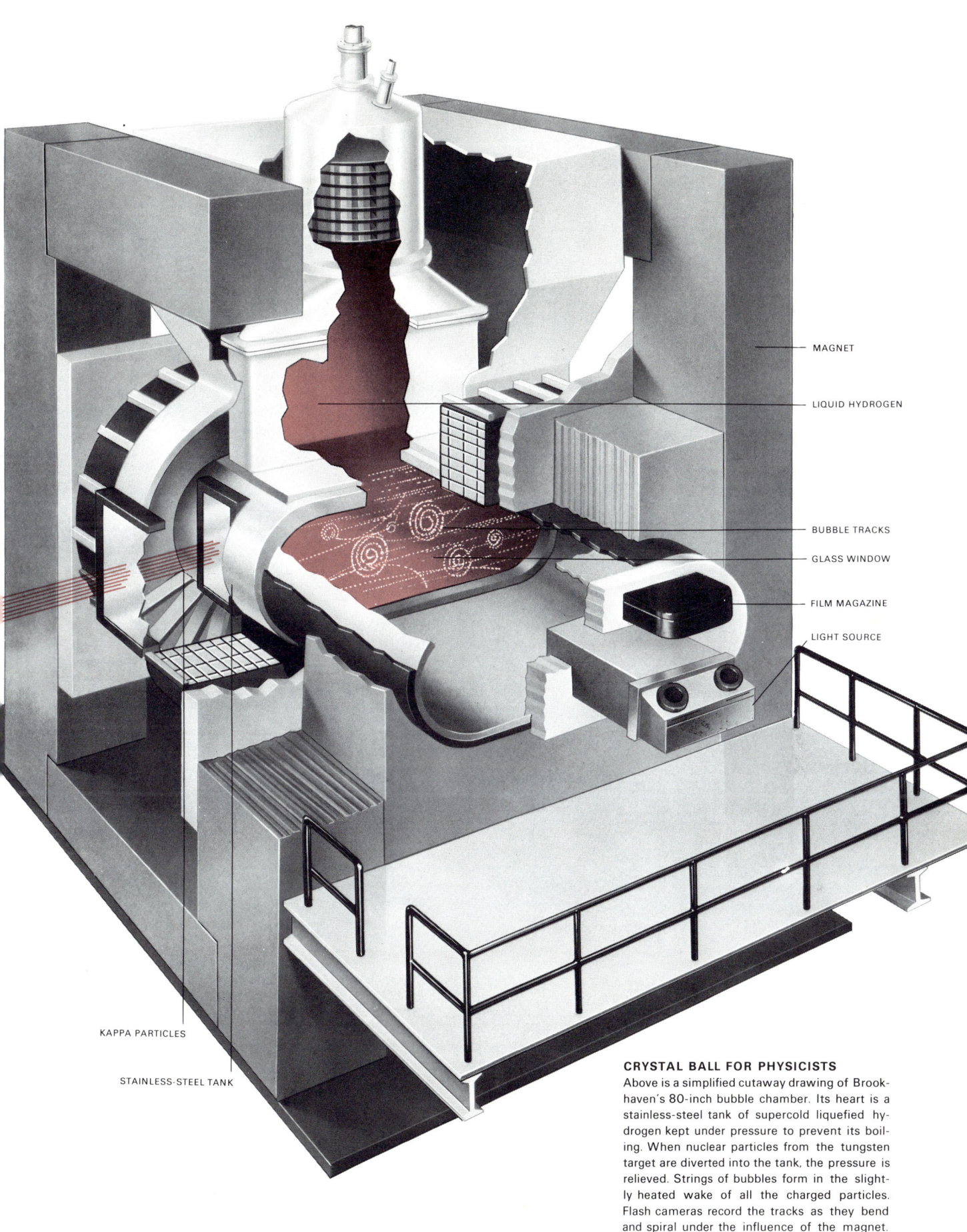

CRYSTAL BALL FOR PHYSICISTS

Above is a simplified cutaway drawing of Brookhaven's 80-inch bubble chamber. Its heart is a stainless-steel tank of supercold liquefied hydrogen kept under pressure to prevent its boiling. When nuclear particles from the tungsten target are diverted into the tank, the pressure is relieved. Strings of bubbles form in the slightly heated wake of all the charged particles. Flash cameras record the tracks as they bend and spiral under the influence of the magnet.

The Atom Smasher's Noisy Trade

The men and machines at Brookhaven turned to the search for Omega Minus in November 1963. With an ear-shattering "wham . . . wham . . . wham," sometimes day and night, the bubble chamber expanded and took its pictures—when it was in good working order. The mammoth but delicate machinery was plagued by vexing disorders at first.

The bubble chamber turned out thousands upon thousands of photographs, each resembling a scratched-up piece of glass—meaningless to the layman but beautifully legible to the technicians, who read meaning in the tiniest scratch.

To create an Omega Minus, Kappa Minus particles of specific energy had to be selected from the mass of particles issuing from the tungsten target. The equipment could deliver to the chamber only about five to 10 of these Kappas per photograph. The physi-

SITUATION NORMAL
During a siege of trouble with the beam of particles, an engineer peers tensely into the oscilloscope that tells him whether Kappa particles are entering the bubble chamber at the proper time and with the proper energy. Periods when everything worked well were relatively rare.

cists' calculations showed that the odds against seeing the trace of an Omega Minus in any one photograph were about one in 50,000.

During November, December and January of 1963 and 1964, however, the tension of the experimenters was not so much over the experimental outcome as over their balky machines. When everything was going right, the shattering "wham . . . wham . . . wham" was a welcome sound.

Closing in on the Quarry

Each time the bubble chamber expanded, three cameras recorded an all-round view of the events inside. As it happened, the experimenters took nearly 300,000 photographs, not counting 75,000 test shots, before they found one showing the career of an Omega Minus particle. Each photograph had to be meticulously examined, for if one tiny scratchlike mark were overlooked, the loss could not be measured even in terms of millions of dollars of wasted manpower and equipment: it would jeopardize a theory whose confirmation was eagerly awaited by physicists around the world.

The deluge of photographs was routed to viewing rooms where specially trained girls looked them over, measuring lengths and angles of all tracks that looked unusual.

On Friday, January 27, 1964, Samios himself was in the scanning room looking over photographs that had appeared intriguing to the girls. He spotted the Omega Minus' telltale mark *(white box)* on the photograph shown below, taken during the 97,025th expansion of the chamber. The tiny track traced the particle's brief career. After only a 10-billionth of a second, it decayed into a negatively charged pi particle, which turned abruptly downward, and a neutral xi particle, whose track was invisible.

OMEGA'S PHOTO FINISH
The three photographs at right show, first, the bubble-chamber log on January 27. At 22:50 (10:50 p.m.), after the 97,089th expansion, the cameras were stopped to insert new film. Experimenters were unaware they had already photographed the "event" they sought about 10 minutes earlier. The next picture is the crucial photograph showing the one-inch track made by the newly created Omega Minus before it decayed. The other curving lines are tracks of other types of particles. At far right is the area within the white box enlarged to about actual size. The arrow points to the track.

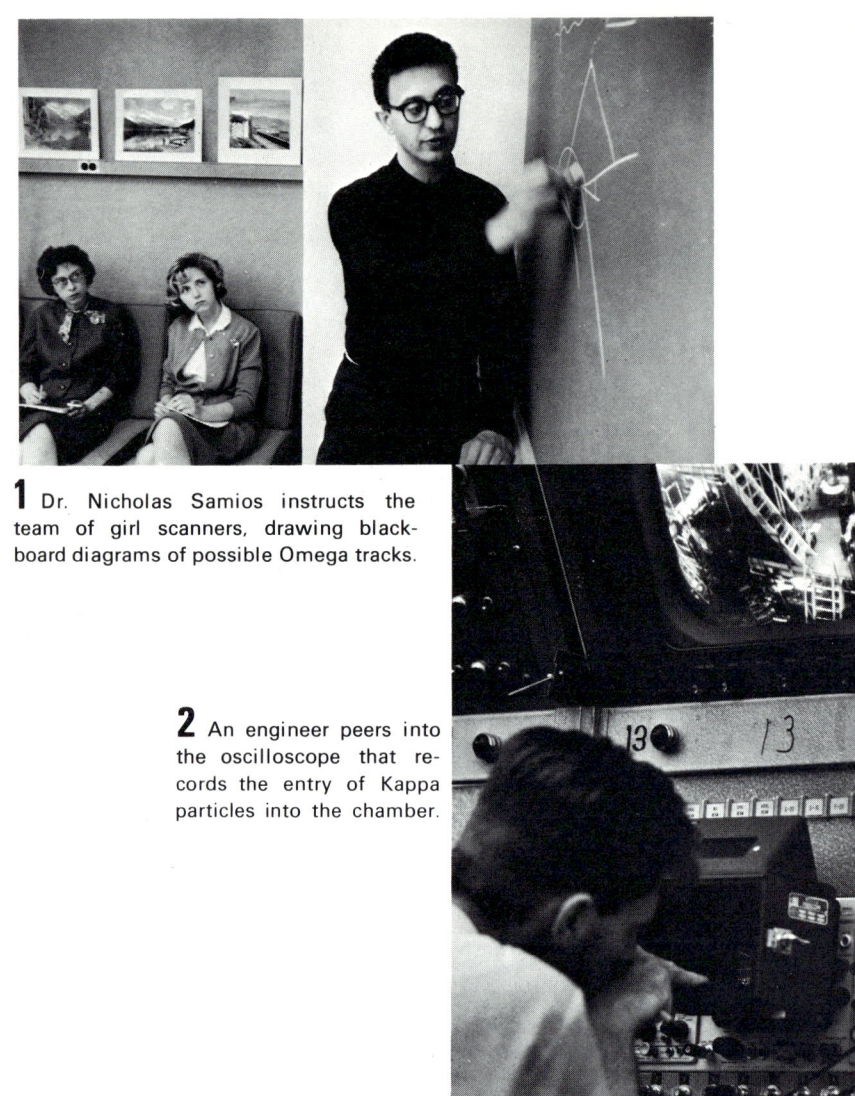

1 Dr. Nicholas Samios instructs the team of girl scanners, drawing blackboard diagrams of possible Omega tracks.

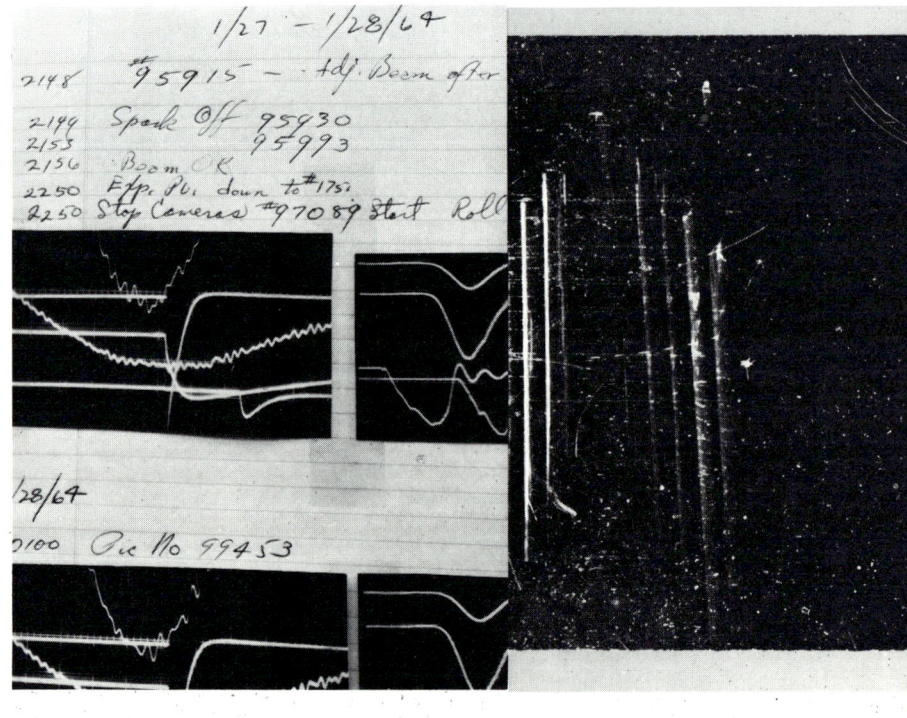

2 An engineer peers into the oscilloscope that records the entry of Kappa particles into the chamber.

3 The chief bubble-chamber engineer on duty superintends the squad of technicians who tend the chamber's controls and its own computer system.

4 Behind technicians working on the bubble chamber is the housing for its photo lights and film magazine.

5 In the physics building, Dr. Samios confers with a scanner over some doubtful photographic tracks.

6 A protractor measures angles of particle tracks.

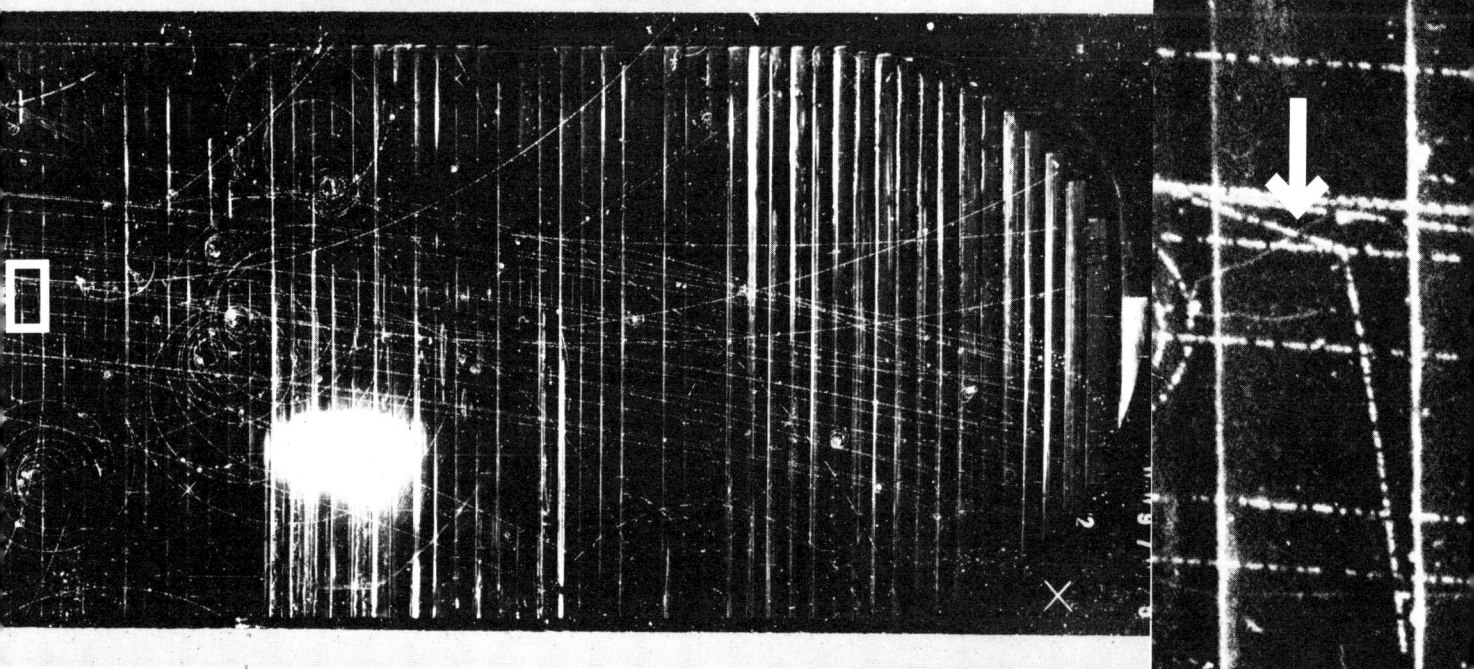

A Still-greater Hunt Ahead

Because of the experimental proof that the eightfold way is an accurate glimpse of atomic order, some elated physicists have equated the discovery of Omega Minus with other classical experiments, like the discovery of nuclear fission. Some feel, furthermore, that this knowledge is a first step toward a "field theory" that will unravel the secret of the relationships of the weak and strong nuclear forces, electromagnetism and gravity. Einstein spent years trying to derive a unified field theory. Says Gell-Mann of the search ahead, "This is the greatest adventure of our time."

Present-day experiments indicate that the origins of these pervasive forces may not be found until accelerators are built with far higher energies than the AGS. The Russians are building a machine to yield 70 billion electron volts (Bev), compared to the AGS's 33 billion, but many physicists believe that energies of at least 800 Bev would be the most revealing. An 800-Bev machine would cost about a billion dollars to build, but in the opinion of the high-energy physicists, the results would be worth it.

SPREADING THE WORD
Gell-Mann, using slides, outlines his theory, and the implications of the Omega Minus discovery, before members of the American Physical Society at Washington's Hotel Shoreham in the spring of 1964. Armed with verification of the eightfold way, physicists are now predicting —and confirming—the existence of still other atomic particles and their various combinations.

A HORDE OF HUNTERS
Posed around and on the Brookhaven bubble chamber are the 114 people who shared in the hunt for Omega Minus. They include 26 physicists, 45 mechanical and electrical technicians, 18 scanners, 14 electrical and mechanical engineers, 11 beam crewmen and two programmers.

4
An Expanding Realm

TODAY A VERITABLE SUPERMARKET OF SPECIALTIES awaits the student contemplating a scientific career. At last count, the number of branches of science listed by the National Science Foundation had reached 620, most of them still unheard of by the average citizen. There is, for example, mesometeorology, which is concerned with major weather phenomena, such as cyclones and tornadoes. There is intuitionism, a specific type of mathematical logic. Dielectrics (not to be confused with either the dialectics of logic or the dialectology of language) deals with the influence of electrical fields on nonconductors. The cryogenicist and the cryologist both derive their names from the Greek *kryos*, "icy cold," but have quite different callings. The one is a physicist who investigates the properties of matter at low temperatures; the other is an earth scientist who concentrates on ice packs and glaciers.

Fortunately, a certain basic order reigns over this seeming chaos. All scientific specialties lie within just four main provinces: mathematics, which has to do with the relationships between numbers, shapes and other logical symbols; the physical sciences, which deal with the inanimate constituents of the universe; the life or biological sciences, which deal with living matter; and the social sciences, which are concerned with human conduct, collective as well as individual. None of these realms is a closed preserve. Mathematics is constantly used in all the others, and a trend to give-and-take is evident in the rise of the so-called "interdisciplinary" sciences. One newcomer, for instance, is bioastronautics, a hybrid of biology, astronautics and space physics, which has to do with the effects of space travel on the human body.

Scarcely a year passes without the appearance of more jaw-breaking "autics," "amics" and "ologies." Scientists continue indefatigably to fence off their special plots in the landscape of general human knowledge. The accelerating pace of this activity fills the nonscientist with a mixture of awe and foreboding. Not surprisingly, he sometimes wonders whether anything will be left in the end—either to learn in school or to ponder with pleasure and profit at home—that has not been scientifically collected, organized and verified.

No way exists either to allay or confirm these apprehensions. In theory, almost any kind of knowledge might be made scientific, since by definition a branch of knowledge becomes a science when it is pursued in the spirit of the scientific method described in the previous chapter. In practice, it is hard to imagine a student of ethics measuring precisely what is best, in terms of individual behavior, for the whole human race; or an artist formulating a theory of visual stimulation that would enable him to evoke predictable reactions from the spectator. A skilled hunter possesses intuitions about wild animals that are not understood in terms of present zoology. The green thumb of the farmer eludes complete botani-

A ROYAL LOOK AT SCIENCE
In this 17th Century engraving, Louis XIV (in plumed hat) enjoys a guided tour of his new *Académie des Sciences* in Paris. Reflecting the era's rapid growth of science, the profusion of equipment displayed for the edification of the king includes a physicist's concave burning mirror, an astronomer's armillary sphere, a cartographer's map, and animal and human skeletons.

A COMET PASSING EARTH was once considered an unlucky omen. One streaked across the heavens on the eve of the Norman conquest of England in 1066, and the medieval Bayeux Tapestry, which relates the story of the conquest, portrays Englishmen pointing to the phenomenon with dread *(above)*. The comet was later named for the English astronomer Edmund Halley, who published a study of comets in 1705. The table below, recording the paths of known comets, was in the book. From it, Halley deduced that the comets of 1531, 1607 and 1682 were the same, and predicted correctly it would reappear in 1758. It will appear next in 1986.

cal explanation. The chemist can analyze a good wine, but not, as yet, a great wine.

Furthermore, not all the subjects practiced as sciences have proved susceptible to full treatment by the scientific method. For instance, paleontology, which is concerned with life of the past as inferred from fossils, does not lend itself to systematic experimentation. Neither does archeology, which is concerned with men of the past as inferred from excavated artifacts. Yet paleontology and archeology are both universally recognized as sciences, while considerable debate rages over the status of history per se, which is also concerned with the past. Some scholars regard it as one of the social sciences; others classify it as one of the humanities. What seems to set it apart from the other two subjects is the nature of the evidence on which it is based. Archeology and paleontology are built at firsthand on artifacts and bones unearthed from the ground and preserved for re-examination. History, by contrast, is built largely at secondhand on the testimony of writers, some of whom may have been liars.

Even among subjects unanimously acknowledged as scientific, some are considered less "scientific" than others. In the pecking order that prevails, a "descriptive" science, one in which the scientist describes, classifies and organizes the data he has collected and examined, is outranked by a "theoretical" science, in which he not only describes his data but works out theories to explain them. A theoretical science, in turn, is outranked by an "exact" science, in which the theories are couched in mathematical terms. The scientist is thereby able to foretell future occurrences, so that his science becomes "predictive" to boot.

An adjustable pecking order

Like most pecking orders, this one is subject to challenge and change. Archeology and paleontology are usually labeled descriptive sciences. So are geology, oceanography, biology, meteorology, anatomy, pathology, psychology and sociology. Yet in all these sciences some areas are becoming almost as theoretical as the exact sciences of physics, chemistry and astronomy. Biologists and meteorologists, while dealing in data far less mathematical than their brethren in such branches of physics as optics and thermodynamics, do not steer clear of prediction. The difference is one of degree. A biologist, given present facts, can only speculate as to future evolution on earth. A meteorologist can tell what the weather will be like a few days hence, but only in terms of probabilities, hedged about by "ifs" and "buts." A physicist, on the other hand, can often foretell the precise outcome of a nuclear reaction even before testing it in a bevatron. An astronomer can calculate to the second the next dozen or more eclipses of the sun. The power to predict accurately hinges on the extent to which the factual raw materials—the protocols discussed in the previous chapter—can be measured and related to one another in terms of mathematics.

Most of the "ologies" of today are merely new subdivisions in provinces of thought more ancient than Gaul. The four mainstreams of scientific interest were in evidence as long ago as the Stone Age. When man learned to count, to know the signs of the seasons, to distinguish nourishing from poisonous plants, and to prescribe a code of conduct for his tribe, he was already engaged in mathematics, physical science, biological science and social science. What has been added to the practice of science is precision and depth of detail. By using the scientific method, the scientist has carved out his special holdings so that they have come to differ from the surrounding terrain: archeology as distinct from antiquarianism, ballistics from marksmanship, botany from gardening.

Requiem for a golden chariot

So impressive is the quality of scientific knowledge that when it has collided with other types of lore, the others have had to yield. Thus the beautiful Homeric conception of the sun as a golden chariot, driven across the sky by a shining young god named Helios, gave way in later Greek times to a theory of the sun circling the sphere of the earth, and then, in the 15th Century, to a theory of the sun standing still while the earth turns.

Historically, men have pursued the quest for scientific knowledge at three different levels of sophistication. Pragmatic experimentation—discovery by trial and error—has a continuous tradition stretching back to the Stone Age. Logical analysis and proof, by careful reasoning from premises, has also been employed since quite ancient times, but the Greeks are usually credited as the first to realize its full potential. The complete scientific method, combining systematic experimentation with analysis and proof, has been used consistently only since the 16th Century. The enormous importance of the second and third levels as watermarks in the rise of Western science was once noted by Albert Einstein in what must be the most succinct history of its sort ever penned. Replying to an inquiry from an admirer, he wrote: "Dear Sir, Development of Western Science is based on two great achievements, the invention of the formal logical system (in Euclidean geometry) by the Greek philosophers, and the discovery of the possibility to find out causal relationship by systematic experiment (Renaissance)."

Pragmatic science flourished in the earliest communities in the valleys of the Indus, the Nile, and the Tigris and Euphrates. In the millennia before the birth of Christ, inventive men conferred on civilization such gifts as the plow, the potter's wheel, glass, the sailboat, the copper foundry and the calendar. The basic urge to know, as well as each new exigency of daily life, evoked a response, and thereby produced the kernel of some science to come. The need for skills in building and in weaponry was to lead to the earliest engineering and physics; the hankering for sturdy tools, to metallurgy and chemistry; the compassion for an injured fellow being, to anatomy and biology. Watching the sun, moon and stars to mark the shifting seasons, early men laid the foundations of astronomy. Measuring time and land, trading with each other, they developed a quickness with numbers which was to give rise to all the quantitative aspects of science.

The Hittite metallurgists who experimented with iron smelting, the Mesopotamian agronomists who devised irrigation systems for the desert wastes beyond the riverbanks, the Chaldean engineers who built tombs and palaces, the Indian and Phoenician merchants who worked out number systems for their accounts and records, the Egyptian military doctors who performed feats of surgery and bonesetting, the old women everywhere who delivered babies and prescribed brews—all were scientists, in a sense. They thought logically, they experimented, they created a technology. From their discoveries flowed the legend of the golden Promethean days when man first became aware of himself as a rational being.

The patterns of revelation

In a few areas of ancient science, particularly arithmetic and astronomy, enough facts were collected, or their repetition in nature was observed often enough, so that they began to reveal patterns, and thus to suggest the possibility of theories. As a result, these subjects were taken up by the priestly castes of Mesopotamia and Egypt, and for the first time, sciences were practiced in scholarly fashion for their own sake. The wise men of Sumer and Babylon, for example, having taught themselves to solve problems in arithmetic and algebra, were able to predict eclipses of the sun and moon. Their exactness was uncanny and so was their method, because it apparently consisted of numerical analysis similar to the work now performed by computers. By taking repeated fixes on the heavenly bodies, century after wearisome century, they compiled long lists of times and positions, and saw in the sequence of numbers certain recurrent themes. From these rhythms they found that they could foretell celestial events far in the future.

As writing was developed and improved, the discoveries of one generation could be handed down to the next, on cuneiform tablets or papyrus scrolls. New vistas opened for the expansion of scientific subject matter, some sound and some not. A considerable body of valid theory concerning arithmetic, algebra and astronomy was passed along by the scientist-

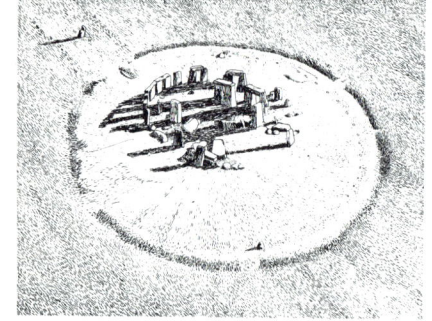

A PREHISTORIC OBSERVATORY, Stonehenge was built on the Salisbury Plain of England between about 2000 and 1500 B.C. From the massive stones and other landmarks that exist today *(above)*, archeologists have deduced the pattern of the original site *(below)* and have long puzzled over its significance. An American astronomer, Dr. Gerald S. Hawkins, recently showed with the aid of computers that the stones were aligned so as to indicate the solstices and the beginnings of seasons, and to predict eclipses of the sun and moon. The arrows drawn below show the alignment of landmarks (stones, pits and the center of the circles) that pointed to the rising and setting of the sun on the days of the summer and winter solstices.

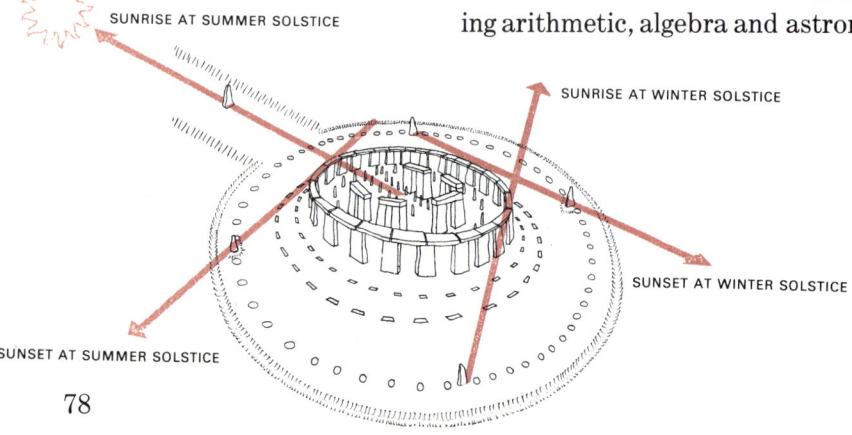

priests, together with a number of false notions, garnished liberally with myths for the masses. These wayward children of early science have come down to us in such "black arts" as Babylonian astrology, Egyptian alchemy, Chaldean necromancy and Etruscan divination. Of these, astrology, in particular, continues to exert such wide fascination that columns devoted to it take up more space in newspapers today than news of astronomy.

When the Greeks inherited the science of the Near East, they brought to it a viewpoint that made possible a tremendous upsurge in scientific growth. In essence, the Greeks held that all nature might be fathomed by methods of deductive logic. Heretofore, scientific theorizing based on observable facts had been, at best, a random affair; now philosophers like Pythagoras and Plato consciously formulated a technique for the would-be theorist. First, he must clearly state his evidence and the assumptions he drew therefrom. Second, he must show, by meticulous step-by-step reasoning, how he arrived at his conclusions.

The chief area in which the Greeks demonstrated this mental tool was geometry, their own special contribution to mathematical science. Yet deduction was potentially applicable to any branch of knowledge. Plato's student Aristotle sought to apply it to biology by experiments in embryology. Stargazers applied it to astronomy, and conceived geometrical models of the solar system which would explain the recurrent patterns of numbers discovered by the Babylonians.

Through the impact of Greek thought, science began to be divorced from soothsaying, and a tradition of free inquiry was launched. Out of the schools of the Athenian philosophers arose the first institutes for theoretical and experimental research.

A place for the muses

The most outstanding of them was founded at Alexandria in Egypt about 300 B.C. by Alexander the Great's general, Ptolemy I, originator of the dynasty which was later to be ruled by Cleopatra. This institution was called a Museum, not in today's sense of the word but as a place dedicated to the Muses—the Greek deities who presided over art and learning. Within its stately walls a small, elite band of scholars, both Greeks from the mainland as well as Hellenized expatriates from Asia Minor, devoted themselves to an intensive and almost completely objective pursuit of science in the modern spirit—pragmatism, deduction and experimentation.

Like the leading research institutes of today, the Alexandria Museum was neither a university nor a laboratory, but seems to have been a combination of both. It awarded no degrees and held no examinations, although novices were required to complete a course of study in theoretical science, the *Logikon,* before they went on to the *Cheirourgikon,* or experimental laboratory. The facilities offered for original research probably included an observatory, botanical and zoological gardens, dissection rooms and special rooms for experiments in physiology. There was also a library which in itself was a major wonder of the times, containing upward of 100,000 papyrus scrolls. Some of this vast collection was obtained by purchase, some by outright expropriation from all over the civilized world. Indeed, the Alexandrian bibliophiles had such a thirst for acquisition that by royal Ptolemaic decree travelers reaching the city with scrolls in their possession were forced to yield them up. If not already available in the library, the scrolls were kept and copies were given the owners.

Amid these unique surroundings shone a succession of luminaries whose works attest to the wide range of subject matter encompassed by science at that early time. Euclid, in his renowned *Elements,* gathered together past mathematical knowledge and added to it. Hero, 1,600 years before James Watt, constructed a miniature steam turbine. Eratosthenes, 1,800 years before Magellan, measured the circumference of the earth by astronomical surveying techniques. Aristarchus, 1,700 years before Copernicus, declared the sun to be the center of the solar system. Herophilus dissected cadavers and described the veins, arteries, nerves, eyes and many internal organs of the body.

Extinction and enrichment

The creativity of the Alexandrian institute languished during Roman times and in the end was extinguished when, after several sackings and burnings, the city fell to the forces of Islam in 642 A.D. Thereafter, the academic pursuit of science passed to Moslem and Byzantine savants, who enriched it by a variety of contributions in algebra, mechanics, optics and medicine. Over the centuries medieval translators and encyclopedists preserved the great scientific heritage of the Greeks which might otherwise have been lost. Initially the Greek works were translated into Arabic; later they were retranslated into Latin. The task, by any standard, was a formidable one. The Arab translators, for example, had to compile an entire Greek-Arabic scientific dictionary to guide their efforts. The hunt for suitable equivalents, however, brought added reward: the Arabs were able to clarify many concepts which the original authors had expressed only in vague terms.

The artisan-scientists of medieval Europe, like their colleagues from Stone Age times onward, continued to augment man's technical knowledge. They contrived more efficient harnesses for draft horses, built improved dikes, harnessed power from windmills and waterwheels, experimented with new ship rigs and hull planking, devised cannons to exploit the

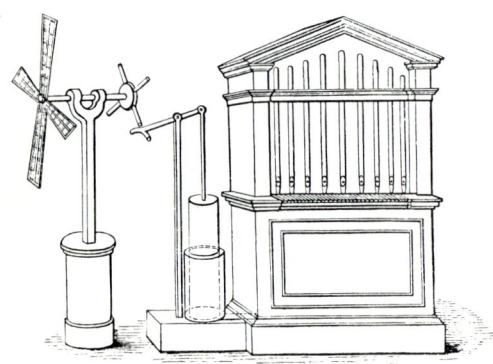

A WIND-POWERED ORGAN was an invention of Hero, a Second Century scientist at the Museum in Alexandria. This drawing, reconstructed from Hero's texts, shows that his organ was an early application of a primary machine principle: the conversion of rotary motion into up-and-down reciprocating motion. A windmill rotated pegs, which bore down on a lever as they turned, causing a piston to rise. When the pegs slipped past, the piston fell and blew air into the organ.

magic of gunpowder, and finally, after the invention of printing, began to broadcast their discoveries in handbooks published in everyday vernacular.

The printing press ushered in a new age of science just as the development of writing had done some 3,000 years earlier. Written records had made it possible to establish a continuous tradition of scholarship. Now the inexpensive mass reproduction of scholarly works made it possible for impecunious craftsmen and inventors to share in the swim of academic science, to read up on theoretical problems, and to exploit fully their skills and ingenuity. In the space of a century, from 1500 to 1600, science as taught in the universities was infused with a fresh spirit rising out of the crafts and trades. The academicians, who had been arguing theories with brilliance and insight for hundreds of years, found their ranks infiltrated by a new type of practical personality, often of lowly birth, whose characteristic attitude was: "Let us cease arguing and find out. Let us experiment."

Spokesman for the future

One of the chief spokesmen of the new age was the Italian Galileo Galilei. More than anyone else, he succeeded in disseminating the belief that, through measurement, the apparatus of mathematics could be fitted to the workings of nature. His most celebrated proof of this consisted in showing that the fall of an object to earth could be described in an equation. His friend and correspondent, the German Johannes Kepler, went even further, devising mathematical laws to match the movements of the planets. And in the 17th Century the Englishman Isaac Newton went further still by fitting all movements of all gravitating objects in the heavens or on earth to a single formula, the law of universal gravitation.

Science as we know it today was born of the wondrous success of induction and experiment, wedded to deduction and mathematics—the success, in short, of the full scientific method. Between roughly 1590 and 1690 a host of geniuses, attracted by its possibilities, produced a flowering of research scarcely equaled in any other 100-year period. Among them, in addition to Galileo, Kepler and Newton, were such giants as Bacon, Gilbert, Boyle, van Leeuwenhoek, Huygens, Descartes, Harvey, Halley and Hooke. No brief summary of their work could do them justice. But the key nature of their role in laying the foundations of modern science may be judged from only a partial list of basic scientific tools associated with their names: the horseshoe magnet, the thermometer, the chronometer, the diverging lens, the reflecting telescope, the compound microscope, the microcaliper, the spring balance and the graph.

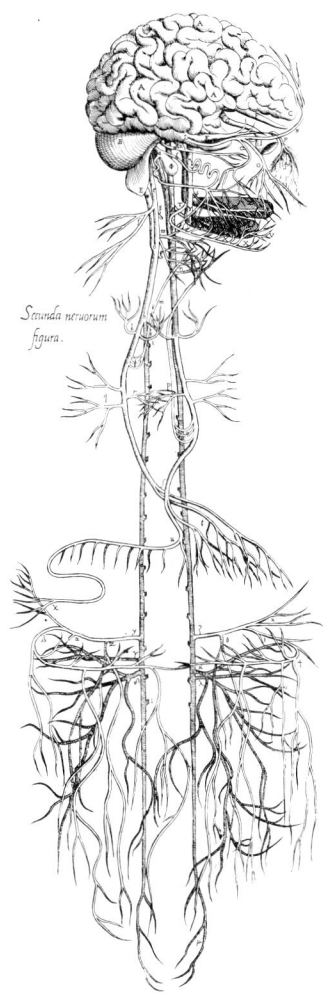

THE SPREAD OF KNOWLEDGE was speeded when 16th Century publishers began to illustrate texts with clear, detailed copper plate engravings like this one showing the human nervous system. Books had become widespread after the invention of the printing press in the preceding century, but until durable copper plates replaced woodcuts and hand-painted illuminations, editions of an illustrated book were limited and expensive. This chart, engraved by Thomas Geminus, appeared in an English medical work in 1545.

In its fully concerted form, the scientific method seemed to apply most readily to astronomy and physics. A few visionaries like Bacon and Descartes saw that it might ultimately apply to all knowledge, but at first most practitioners of the disciplines we now call sciences went on thinking of themselves simply as scholars. Gradually, however, this viewpoint changed. From the time of Newton, who died in 1723, up to the coining of the word "scientist" by William Whewell in 1840, the various specialists one by one fell under the influence of the scientific vision and came to acknowledge themselves as men of science. As may be seen in the picture essay on pages 84 through 101, the structure of science increasingly expanded to make room for them.

In the 18th Century, Antoine Lavoisier sounded the knell of alchemy and began raising chemistry to the rank of an exact science. Carl von Linné, more widely remembered as Linnaeus, provided the first consistent, detailed and comprehensive classification system for plants and animals. Abraham Gottlob Werner and James Hutton helped transform paleontology, geography and mineralogy into the modern science of geology. Adam Smith brought a novel theoretical approach to the social science of economics.

Curiosity in stout shoes

During the same age of enlightenment, scientific communities reached across national boundaries through exchange students, intensive correspondence, and a multiplication of scientific journals and international societies. More and more, governments sent out expeditions to measure the earth, to chart its wilds and to collect hitherto unfamiliar specimens of its life. A venerable precedent existed for such journeys. Men of scientific bent had gone along with Alexander the Great on his campaigns, observing, among other things, the plant life in the lands he conquered, and bringing back evidence for stay-at-homes to ponder. In the Seventh Century A.D. the Roman scholar Severinus had urged: "Go, my Sons, buy stout shoes, climb the mountains, search . . . the deep recesses of the earth. . . . In this way and in no other will you arrive at a knowledge of the nature and properties of things." This sage advice was followed to such good effect by scientists of the 18th and early 19th Centuries that new collections of facts, and new theories concerning them, have been spawning ever since—too fast for any one eye to follow, for any one book to encompass.

Since the 18th Century, the pursuit of science has become a constant contest between observational scientists, bringing in new loads of facts, and theoretical scientists, seeking to make sense out of the accumulating heaps. During the 19th Century, even as the data-gatherers continued their spectacular advances, the theory-makers began to score significantly as well. It was the century in which Matthias Schleiden and Theodor Schwann proposed that all creatures are composed of cells; in which Louis Pasteur and Robert Koch advanced the idea that most diseases are caused by microorganisms; in which the brothers Grimm, of fairy-tale fame, successfully generalized about the evolution of languages; in which Sigmund Freud began to put forth his theories of the human mind.

It was also the century of Darwin, Maxwell, Dalton and Mendeleyev. In formulating his theory of evolution, Charles Darwin put together the fossil evidence gathered by geologists, the statistics of population surveys, and a vast amount of field knowledge concerning the hierarchy of plants and animals. In concluding that light is a form of electromagnetic energy, James Clerk Maxwell drew on investigations of chemists into the nature of heat, of physicists into the nature of magnetism and electricity, and of mathematicians into differential equations. In popularizing the theory of atomic weights, John Dalton established the idea of the atom as a single kind of building block for all kinds of matter. In his periodic table of elements, Dmitri Mendeleyev suggested to his successors a regularity in the construction of different kinds of atoms which led, in turn, to an appreciation of the most fundamental building modules yet known: the electron, proton and neutron. By the end of the century physicists were beginning to identify specific wavelengths of energy with different types of upheaval in different kinds of atoms.

The effort to amalgamate hordes of particulars into broad theories was intensified in the 20th Century. Einstein, insisting that the laws of physics were simple and invariant, showed that matter and energy are interconvertible. Moreover, in his search for a "unified field theory," he even hinted at the possibility, as yet unproved, that the expansion of the entire universe may somehow be linked to the nature of the atom.

Aspects of an atomic mine

Today, all the physical sciences and some of the life sciences deal in related aspects of a very few basic ideas. Astronomers, biochemists and neurologists all find themselves concerned with particles, atoms, and molecules—with fundamental units moving and combining in a geometric framework of space and time. Indeed, it may some day turn out that all the specialists of science have been working different shafts of the same atomic mine.

In attempting to unify the diverse branches of science, theorists have run into large difficulties. More and more they talk of current scientific problems as involving "complex systems." The "system" may be a living organism, or the world's weather, or a bomb about to be tested, or a nation's economy, or the entire universe. In every case, what is meant is

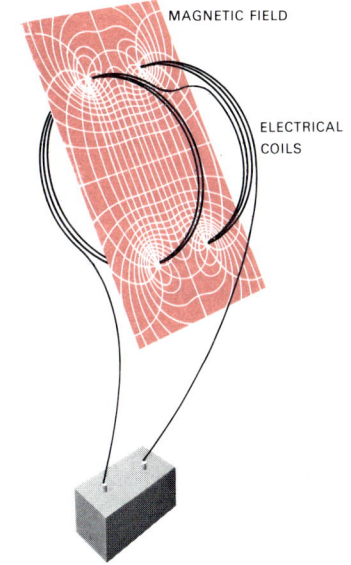

ELECTROMAGNETIC WAVES pulsing through space were first postulated by the theories of James Clerk Maxwell, 19th Century physicist. Before Maxwell's time, it was known that electric currents in two parallel coils create the magnetic field diagramed above. Maxwell demonstrated that when the currents are oscillated, the magnetic field creates a pulsating electric field beyond it, which in turn creates another magnetic field, ad infinitum. He deduced that the speed of these "electromagnetic waves" is the same as the speed of light, and suggested that light itself is an electromagnetic phenomenon. His work led to discovery of the spectrum of electromagnetic waves of different lengths—including radio waves and X-rays.

simply that there are so many variables and so many probabilities—so many individual unpredictable atoms—that no pat theories can be formulated; no sure answers can be attained except through statistics and a weighing of uncertainties. The problems of science have become so complicated that the logical, straightforward equations of the past no longer entirely suffice. To make a start on such problems, mathematicians exploit the lightning arithmetic of computers, which enables them, at least, to calculate approximate solutions.

For the future, it seems reasonable to assume that the increasing use of statistical analysis will extend the theorist's power to deal with complex systems of all sorts. The specialties will probably continue to breed more specialties, and theories will probably continue to strengthen the family ties between them. Not all observers agree on how far the new statistical approach may ultimately go. Some see almost endless possibilities in polling and computing, and believe that these techniques may some day be used to turn all branches of knowledge into statistical sciences. Others say, with equal conviction, that no subject touching on the spirit of man will ever be converted into a set of numbers or a printout from a data-processing machine. Whichever school of prophecy proves correct, it is certain that no end to the spread of science is yet in sight.

The Family Trees of Science

A major aim of science is to reduce diversity to a few general laws. Paradoxically, however, in the process of working toward this unifying end, the sciences have split into ever narrower specialties. There are two usual steps in the creation of a new specialty: (1) someone strikes an unexpectedly rich vein of new knowledge in a corner of one field, and (2) someone in a university department applies a name to it and includes it in the list of courses offered. In this essay, the family trees of the seven major areas of science *(opposite)* are traced as they have branched and rebranched from ancient to modern times. Only the major specialties of each field are shown. Chemistry alone has about 150 subbranches, but only nine are shown. The genealogies graphically raise a question many scientists anxiously ask themselves: how can a man work on one tiny twig of one tree of knowledge and yet stand off from time to time to take a look at the forest?

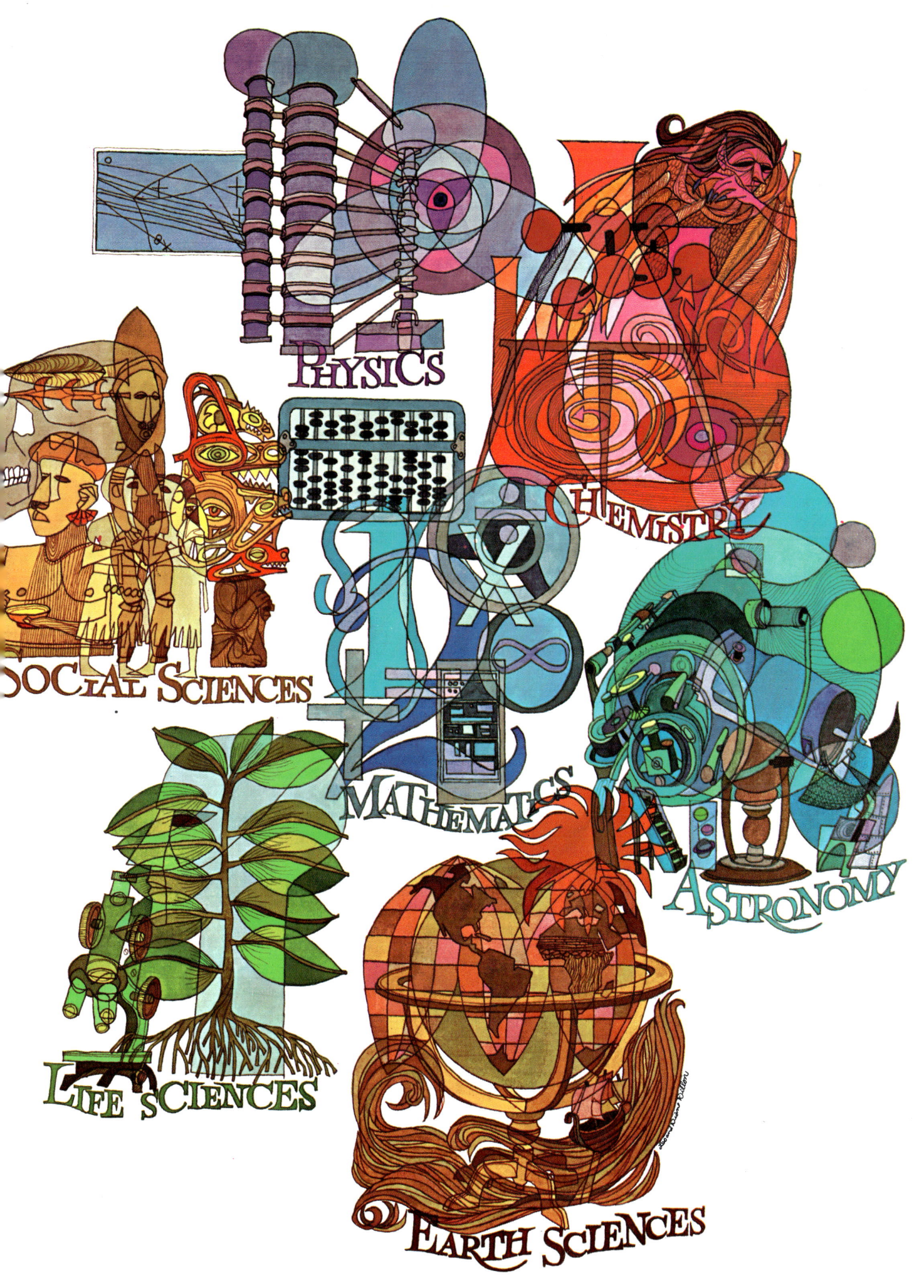

MATH-EMATICS

The Study of Numbers and Shapes

All mathematics has branched from two separate trunks: arithmetic and geometry, or the art of computing and the science of shapes and sizes. The first was used by ancient peoples in record-keeping, the second in construction, surveying and mapping the stars.

The Greeks, with their passion for pure intellectual play, developed the two and fashioned from them number theory, analysis, trigonometry and algebra. They invented a method of thought, logic, and applied it to geometry to create our present system of theorems and proofs.

The 17th Century was another vigorous period for mathematics. Galileo permanently knit mathematics into physics by using geometry to calculate the way a falling body accelerates. Descartes invented analytic geometry by drawing algebra and geometry together into a useful system of graphing. Pascal and Fermat used mathematics to predict the fall of dice, thereby initiating probability theory. Finally, Newton and Leibnitz separately invented the calculus, the highest achievement of this highly productive period.

In the 19th Century a few theorists, such as Bernhard Riemann, building on the work of Karl Gauss, developed non-Euclidean geometries, dealing with imaginary curved spaces and spaces of more than three dimensions. In the present century, information theory has come in time to help program electronic computers.

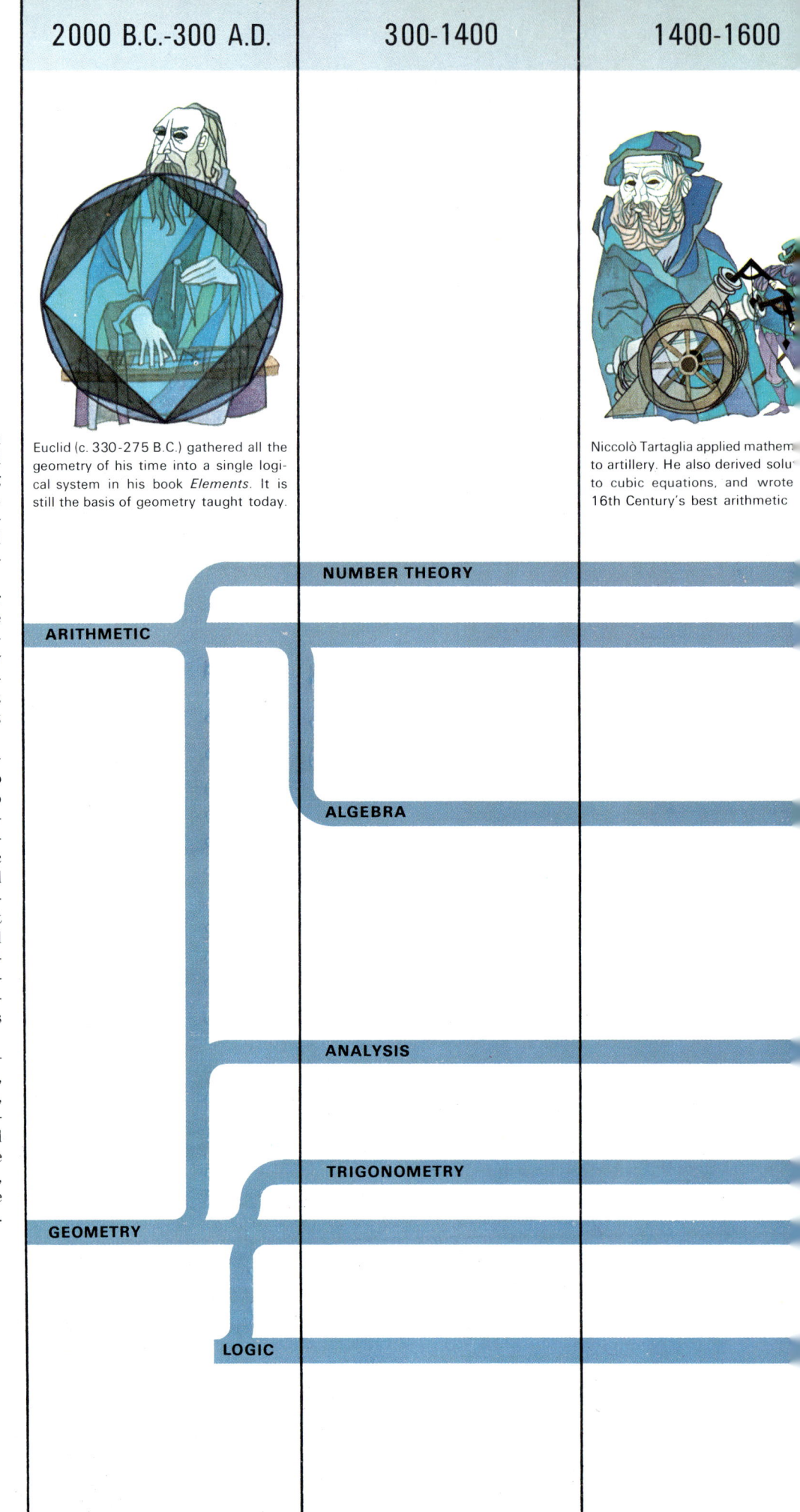

2000 B.C.-300 A.D.

Euclid (c. 330-275 B.C.) gathered all the geometry of his time into a single logical system in his book *Elements*. It is still the basis of geometry taught today.

1400-1600

Niccolò Tartaglia applied mathematics to artillery. He also derived solutions to cubic equations, and wrote the 16th Century's best arithmetic.

| 17TH CENTURY | 18TH CENTURY | 19TH CENTURY | 20TH CENTURY |

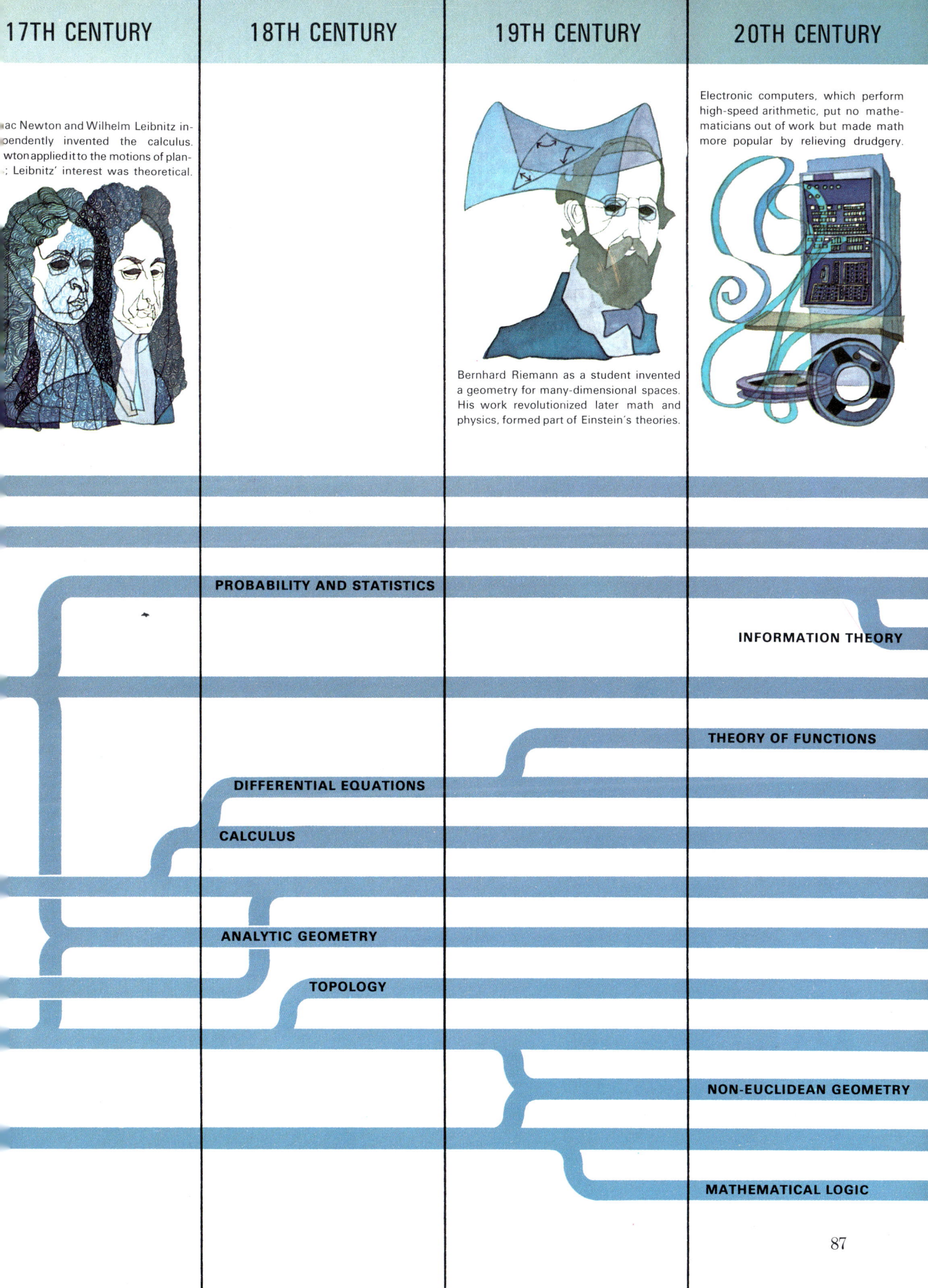

Isaac Newton and Wilhelm Leibnitz independently invented the calculus. Newton applied it to the motions of planets; Leibnitz' interest was theoretical.

Bernhard Riemann as a student invented a geometry for many-dimensional spaces. His work revolutionized later math and physics, formed part of Einstein's theories.

Electronic computers, which perform high-speed arithmetic, put no mathematicians out of work but made math more popular by relieving drudgery.

PROBABILITY AND STATISTICS

INFORMATION THEORY

THEORY OF FUNCTIONS

DIFFERENTIAL EQUATIONS

CALCULUS

ANALYTIC GEOMETRY

TOPOLOGY

NON-EUCLIDEAN GEOMETRY

MATHEMATICAL LOGIC

PHYSICS

The Study of Mass and Energy

Physics, the most basic of the natural sciences, seeks to establish mathematical laws to explain and predict the behavior of mass and energy. Matter had been studied since ancient times, but modern physics suddenly coalesced at the end of the 19th Century out of four apparently unrelated branches. These were mechanics, the study of motion; optics, the study of the properties of light; thermodynamics, the study of heat; and electromagnetism, the study of the properties of electric and magnetic forces.

During the 19th Century, these four fields began to show things in common. Heat was recognized as a manifestation of motion among tiny molecules. Light was found to be an electromagnetic wave. Electromagnetic waves, in turn, were found to behave like certain mechanical systems. These resemblances were explored by some of the best minds in the history of science, men like Maxwell and Planck—and shortly modern "physics" exploded into being.

No other science to date has experienced such heady successes as followed during the next three quarters of a century. The branches of physics proliferated. Atomic physics gave rise to quantum mechanics and to solid-state, molecular and nuclear physics. The last, in turn, spawned particle and plasma physics. Meanwhile, relativity theory and quantum mechanics suggested startling new physical—and philosophical—ideas.

2000 B.C.-300 A.D.

This ornate steam-driven toy is one of many inventions of Hero of Alexandria. Most incorporated important principles of physics never put to significant use.

300-1400

MECHANICS

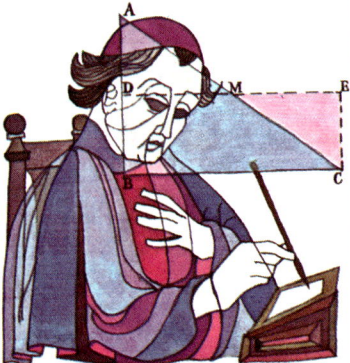

A 14th Century mathematician, Nicole Oresme, used geometry to represent an object's change in velocity in an early application of mathematics to physics.

MINERALOGY

OPTICS

1400-1600

Galileo Galilei applied modern scientific method by using observation, experiments and mathematical theory to establish laws of falling bodies.

88

| 17TH CENTURY | 18TH CENTURY | 19TH CENTURY | 20TH CENTURY |

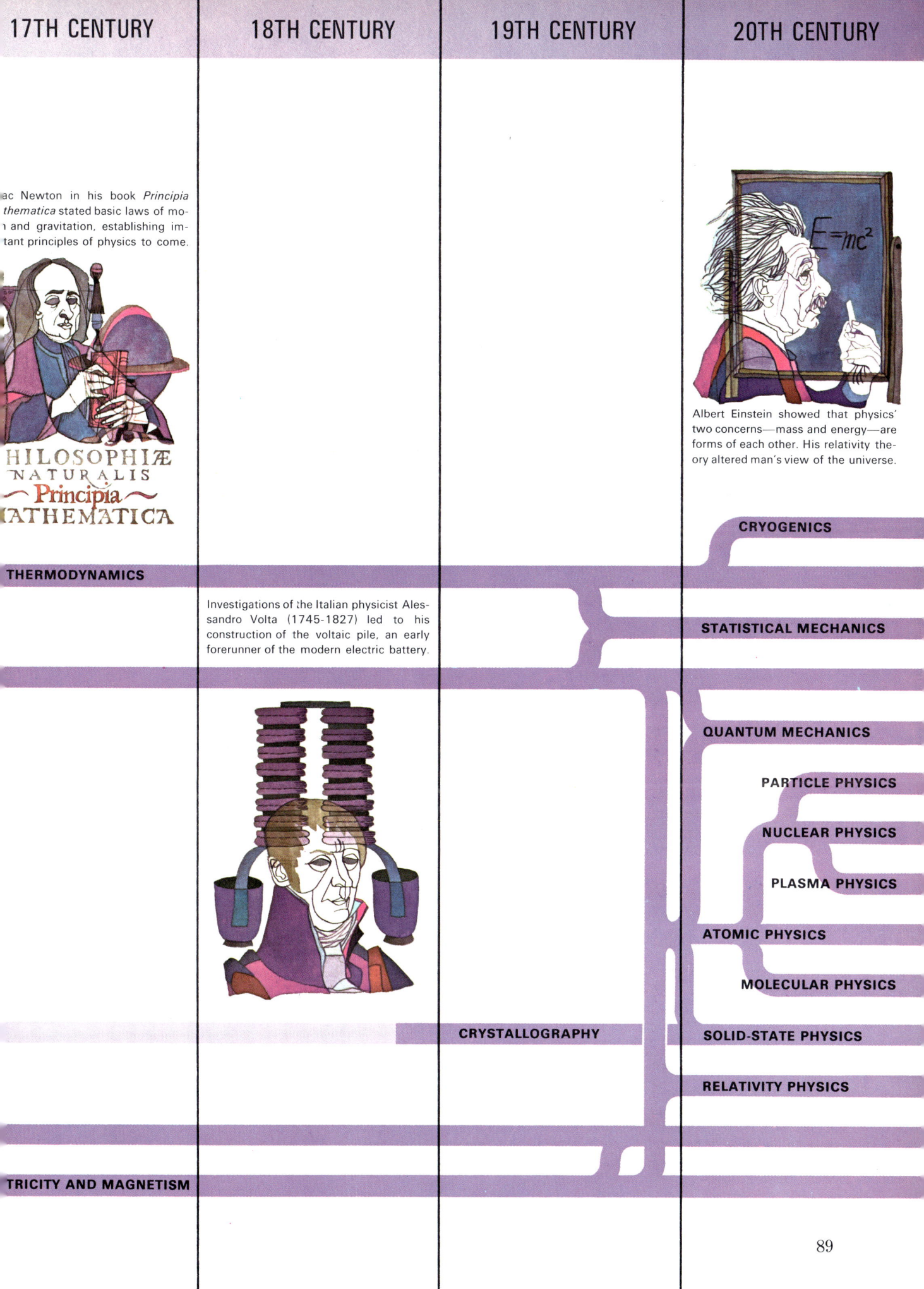

ac Newton in his book *Principia thematica* stated basic laws of mo- n and gravitation, establishing im- tant principles of physics to come.

Albert Einstein showed that physics' two concerns—mass and energy—are forms of each other. His relativity theory altered man's view of the universe.

CRYOGENICS

THERMODYNAMICS

Investigations of the Italian physicist Alessandro Volta (1745-1827) led to his construction of the voltaic pile, an early forerunner of the modern electric battery.

STATISTICAL MECHANICS

QUANTUM MECHANICS

PARTICLE PHYSICS

NUCLEAR PHYSICS

PLASMA PHYSICS

ATOMIC PHYSICS

MOLECULAR PHYSICS

CRYSTALLOGRAPHY   SOLID-STATE PHYSICS

RELATIVITY PHYSICS

TRICITY AND MAGNETISM

CHEMISTRY

| | 2000 B.C.-300 A.D. | 300-1400 | 1400-1600 |

The Study of Substances

Chemistry, the science that investigates the properties and transformations of substances, had its origins in the laboratories of the alchemists. Alchemy (shown as a shaded line at right because its main purpose—to turn base metals into gold—was not the study of chemistry) eventually expired under the impact of science, but its tools and procedures were taken over in the 16th Century by the new study of medical chemistry. Men such as Paracelsus and Van Helmont gained shrewd insight into chemical laws as they attempted to find cures for bodily ailments. Lavoisier's provocative experiments with combustion in the 18th Century completed the transformation of chemistry into an exact science.

Most of the knowledge gathered by these men was in the field of inorganic chemistry. As late as the 19th Century, chemists believed that a mysterious "vital force" was necessary to make substances like those created by living things, the so-called organic compounds, all containing carbon. In 1828, however, Friedrich Wohler produced the organic compound urea in a simple laboratory experiment. Synthetic organic compounds are among our most useful chemical products today. Their recent soaring proliferation has largely been due to the discovery of "polymerization"—a process that enables chemists to create new, tailor-made molecules and link them together to make plastics, textiles or medicines.

In their search for the formula to make gold, medieval alchemists developed many chemical compounds and valuable apparatus like this distillation device.

The earliest chemists were practical craftsmen. This drawing, from an ancient wall painting, shows Egyptians mixing molten metals to make bronze.

CHEMICAL CRAFTS | **ALCHEMY**

Paracelsus, a German physician alchemist, laid the foundations medical chemistry in the 16th C tury by using chemicals as medici

90

| 17TH CENTURY | 18TH CENTURY | 19TH CENTURY | 20TH CENTURY |

Irishman Robert Boyle introduced rigorous scientific method to the study of chemistry, helping rid it of the occult ideas of alchemy in the process.

Apparatus used by Antoine Lavoisier in his precise experiments showed that combustion was a chemical reaction involving air and not the mythical fluid, phlogiston.

Below is the symbol for gold on a periodic chart of elements. First devised by Mendeleyev, the chart lists elements by weight and groups them according to properties.

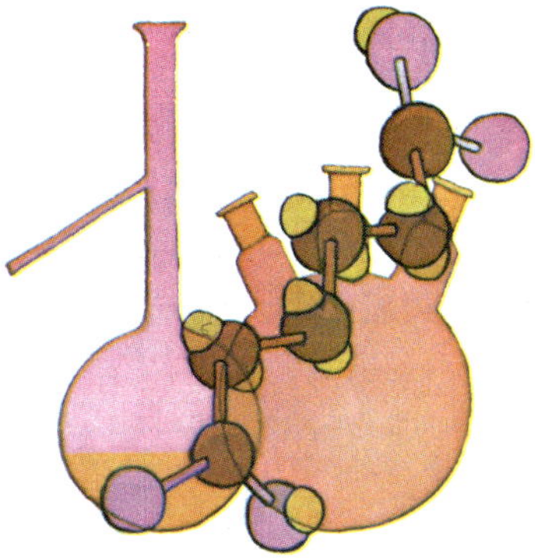

In the 1920s came the discovery of how to "polymerize," or combine, molecules like the nylon 66 component modeled above into long, useful chains.

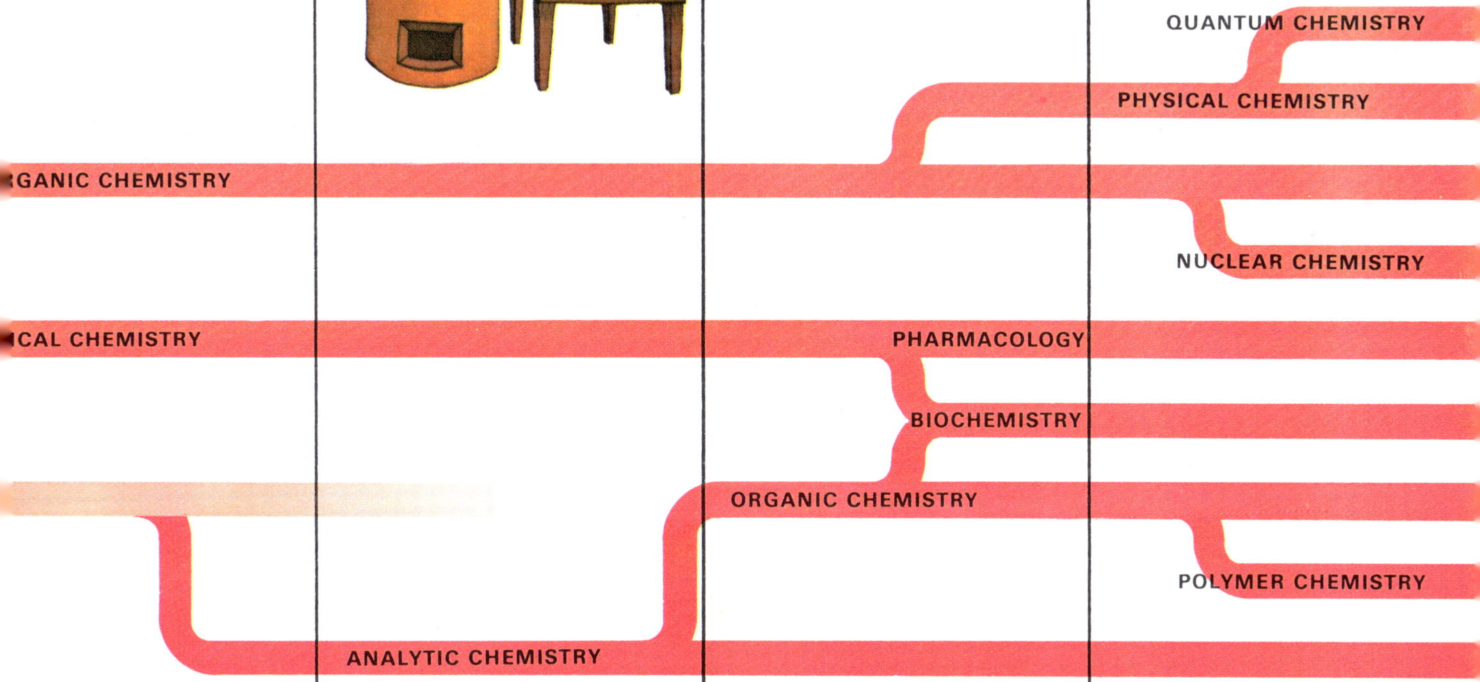

QUANTUM CHEMISTRY

PHYSICAL CHEMISTRY

ORGANIC CHEMISTRY

NUCLEAR CHEMISTRY

MEDICAL CHEMISTRY

PHARMACOLOGY

BIOCHEMISTRY

ORGANIC CHEMISTRY

POLYMER CHEMISTRY

ANALYTIC CHEMISTRY

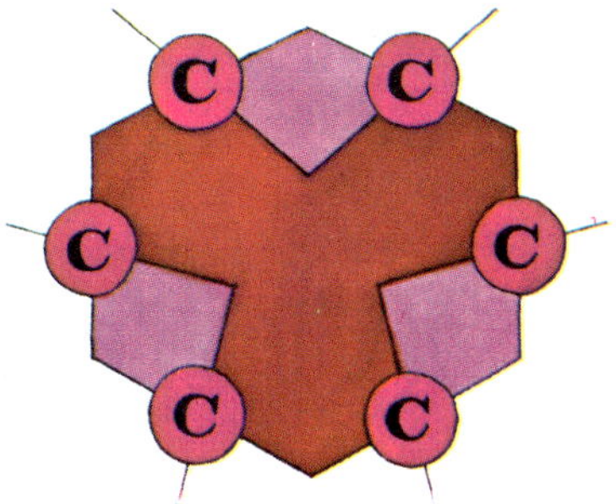

The benzene ring, as conceived by Friedrich Kekule, helped chemists understand the nature of organic molecules, and furthered creation of synthetic compounds.

ASTRONOMY

The Study of Heavenly Bodies

Astronomers originally observed the heavens for the practical purpose of marking seasons, so that such matters as planting could be carried out on time. True to form, the Greeks picked up these beginnings and created cosmology, the theoretical study of the origins and structure of the universe. Soon they combined cosmology and the ancient subject of positional astronomy to form the study of celestial mechanics, or the motions of the heavenly bodies.

Since that time, major astronomical advances have been largely dependent on the development of new instruments. The 17th Century invention of the telescope and the 19th Century spectroscope made possible physical astronomy and astrophysics, which study the character and composition of planets and stars.

After centuries of moving in as stately a fashion as the apparent motions of the heavenly bodies, astronomy has enormously speeded up its pace in the last few decades. One stimulant has been the radio telescope, which has greatly increased the range and value of earthbound observations. The other is space flight, with its urgent need for knowledge and its promise of revolutionary capabilities. Because of the enormous concentration of effort to land men on the moon, for example, information about the moon has probably about doubled in the past five years.

2000 B.C.–300 A.D. | **300–1400** | **1400–1600**

The Second Century Greek astronomer Ptolemy taught that the sun and planets move around the earth. His false ideas dominated astronomy for 14 centuries.

This Copernican diagram shows earth and other planets revolving around the sun. By reversing Ptolemy, Copernicus caused a controversy that raged for over a century.

COSMOLOGY

CELESTIAL MECHANICS

POSITIONAL ASTRONOMY

Arabs kept astronomy alive during the Middle Ages. This drawing taken from a medieval manuscript shows Moslem astronomers recording star positions.

17TH CENTURY	18TH CENTURY	19TH CENTURY	20TH CENTURY

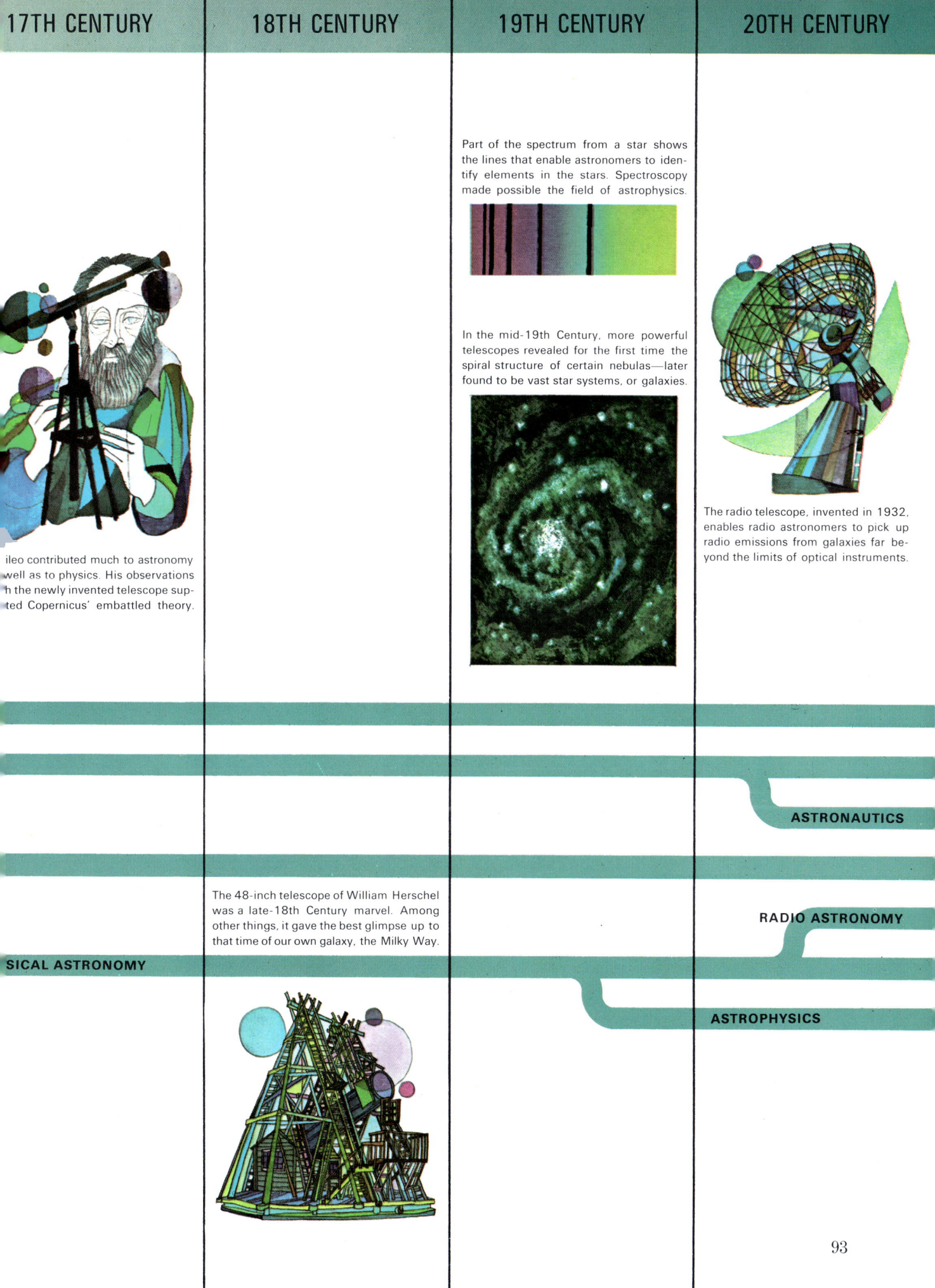

Part of the spectrum from a star shows the lines that enable astronomers to identify elements in the stars. Spectroscopy made possible the field of astrophysics.

In the mid-19th Century, more powerful telescopes revealed for the first time the spiral structure of certain nebulas—later found to be vast star systems, or galaxies.

The radio telescope, invented in 1932, enables radio astronomers to pick up radio emissions from galaxies far beyond the limits of optical instruments.

ileo contributed much to astronomy well as to physics. His observations h the newly invented telescope supted Copernicus' embattled theory.

The 48-inch telescope of William Herschel was a late-18th Century marvel. Among other things, it gave the best glimpse up to that time of our own galaxy, the Milky Way.

SICAL ASTRONOMY

ASTRONAUTICS

RADIO ASTRONOMY

ASTROPHYSICS

EARTH SCIENCES

The Study of Our Planet Home

As the chart on these pages indicates, most of the earth sciences sprang into existence almost simultaneously at the beginning of the 19th Century. Before this specialization, however, several of man's ancient activities had led to some knowledge of the earth. The most fruitful were geodesy (measurements of the earth's size and shape), exploration (a dashed line here because it is not a science), cosmology (a shaded line because it is an astronomical, not an earth science) and mineralogy.

The surge of activity at the close of the 18th Century was inspired by a raging controversy over the age of the earth—a controversy fueled by religious factors. Fossil remains being discovered around the world called into question the Biblical account of ancient history, indicating that the earth was far older than anyone had thought. Men began to investigate the earth scientifically for the first time, freed of ancient presuppositions. The earth sciences split and split again, with separate branches to study the oceans and the composition and history of the earth.

Partly because of this late start, knowledge about the planet we live on is still relatively small and controversial. All the scientific data accumulated up to now, for example, have not explained the earth's origins. It may well be that astronomers or spacemen will discover these secrets before the earth scientists do.

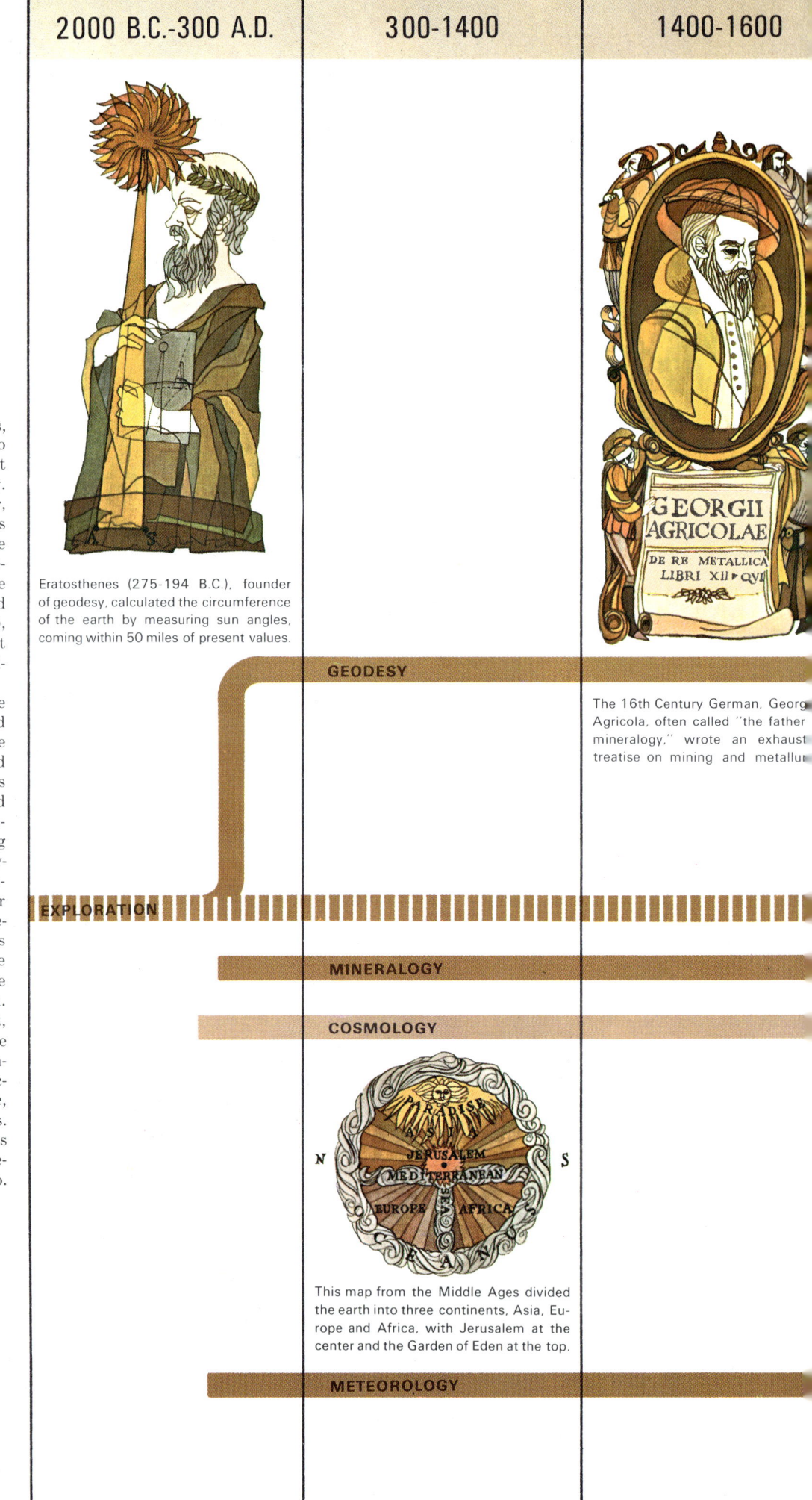

2000 B.C.-300 A.D.

Eratosthenes (275-194 B.C.), founder of geodesy, calculated the circumference of the earth by measuring sun angles, coming within 50 miles of present values.

300-1400

GEODESY

EXPLORATION

MINERALOGY

COSMOLOGY

This map from the Middle Ages divided the earth into three continents, Asia, Europe and Africa, with Jerusalem at the center and the Garden of Eden at the top.

METEOROLOGY

1400-1600

The 16th Century German, Georg Agricola, often called "the father mineralogy," wrote an exhaust treatise on mining and metallu

94

| 17TH CENTURY | 18TH CENTURY | 19TH CENTURY | 20TH CENTURY |

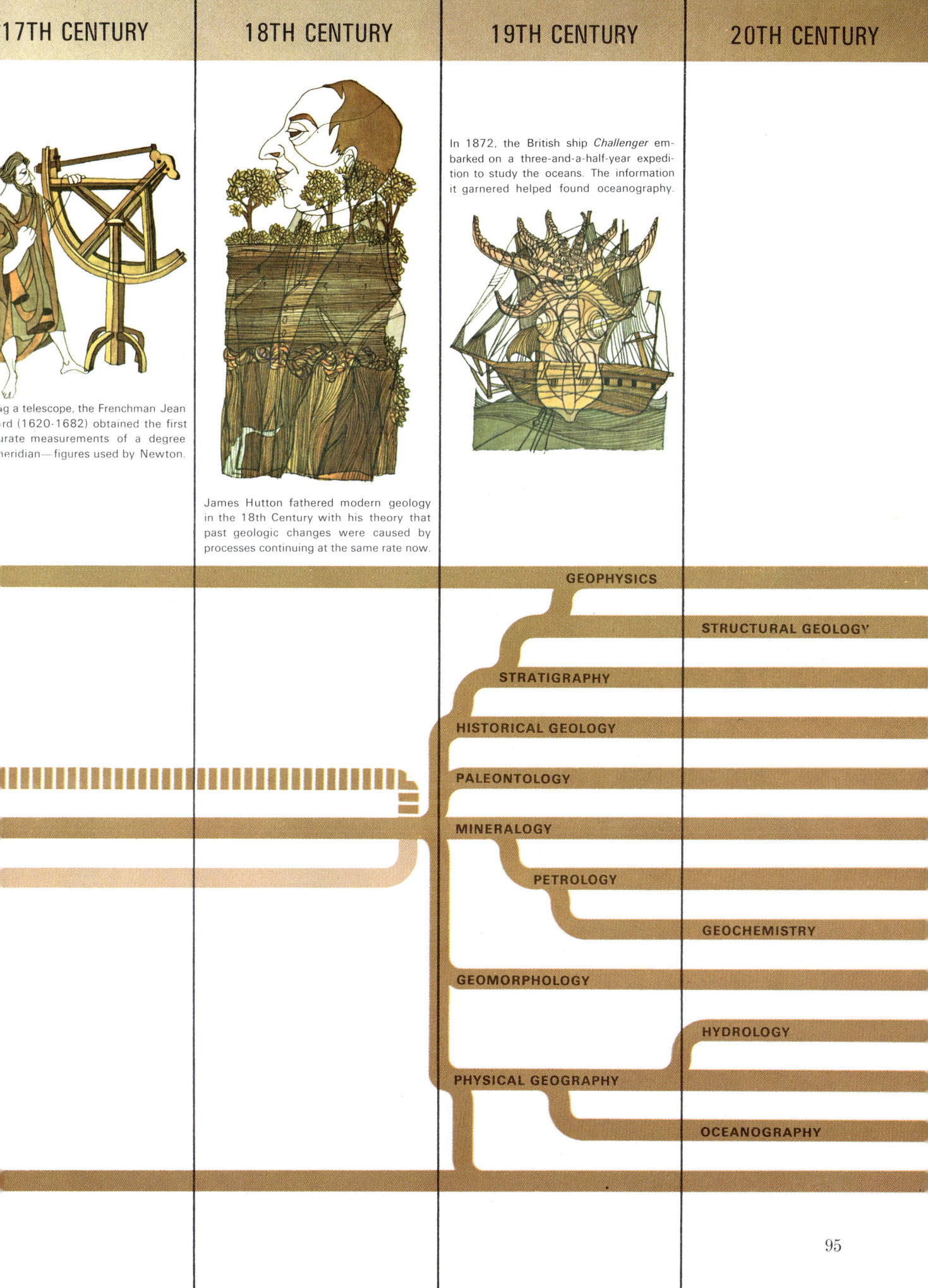

g a telescope, the Frenchman Jean
rd (1620-1682) obtained the first
urate measurements of a degree
eridian—figures used by Newton.

James Hutton fathered modern geology in the 18th Century with his theory that past geologic changes were caused by processes continuing at the same rate now.

In 1872, the British ship *Challenger* embarked on a three-and-a-half-year expedition to study the oceans. The information it garnered helped found oceanography.

- GEOPHYSICS
- STRUCTURAL GEOLOGY
- STRATIGRAPHY
- HISTORICAL GEOLOGY
- PALEONTOLOGY
- MINERALOGY
- PETROLOGY
- GEOCHEMISTRY
- GEOMORPHOLOGY
- HYDROLOGY
- PHYSICAL GEOGRAPHY
- OCEANOGRAPHY

LIFE SCIENCES

The Study of Plants and Animals

The study of living things began with early man's concern for his health. He studied herbs for their medicinal value and learned certain things about his body from his primitive attempts at therapy. The Greeks took this simple body of knowledge and vastly enlarged it, basing much of their medicine on sound natural science. Aristotle's speculations about natural things led him to pioneer in botany, zoology and embryology.

The development of the microscope in the 17th Century led to microbiology, which later gave rise to histology, the study of tissues, and cytology, the study of cells.

As a result of recent discoveries in cytology, an undercurrent of excitement now prevails in university biology departments across the country, recalling the mood of physics during the eventful 60 years just past. The molecular structure of DNA, the substance in cell nuclei that determines genetic characteristics of all living creatures, has been analyzed, and the processes by which it governs cell specialization are becoming better understood. The study of DNA and its associated molecules has recently given rise to the field of molecular biology.

Even middle-aged physicists and chemists are admitting that if they were graduate students choosing a speciality again, they might pick the exciting field of molecular biology.

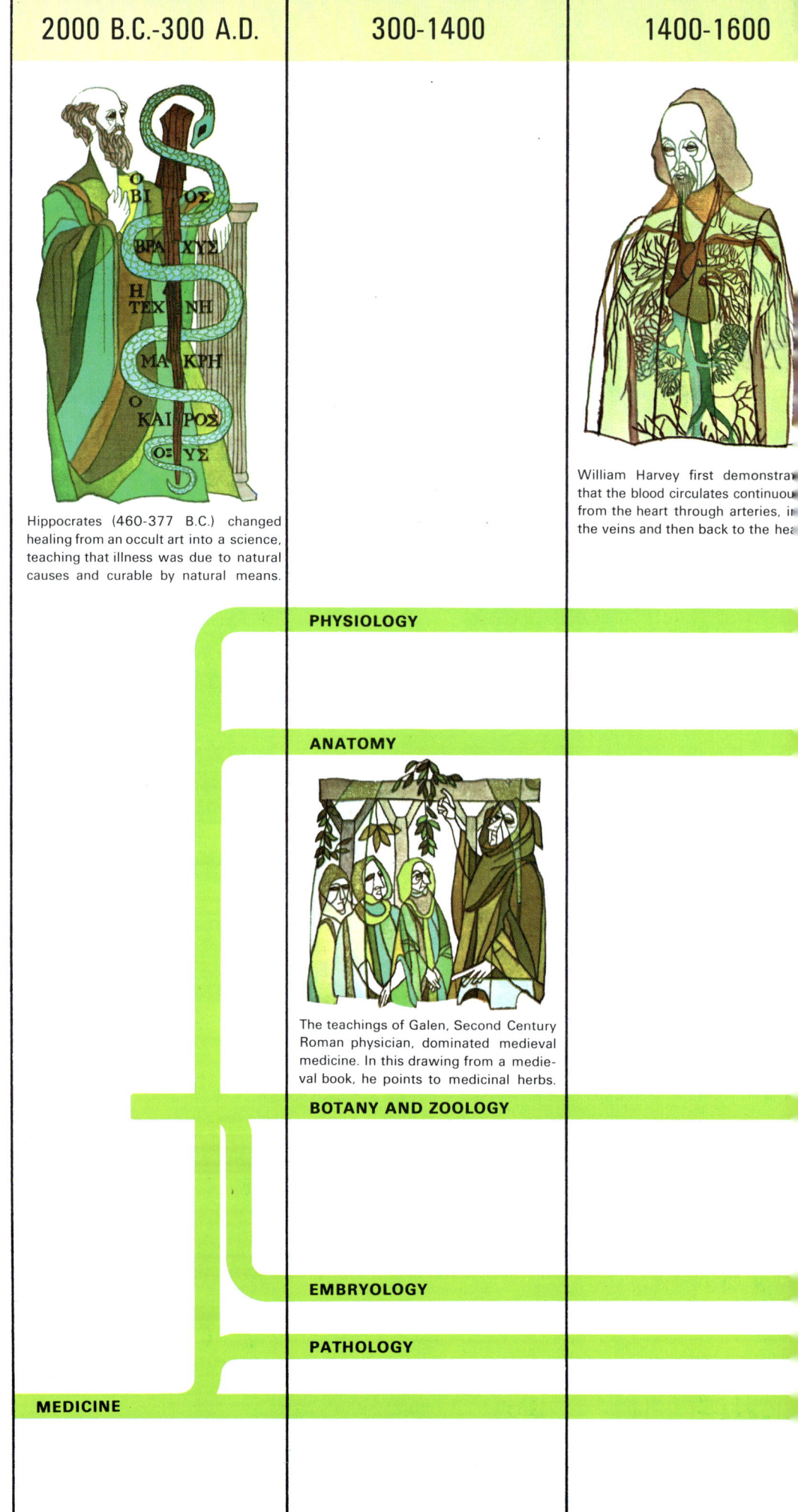

2000 B.C.-300 A.D.

Hippocrates (460-377 B.C.) changed healing from an occult art into a science, teaching that illness was due to natural causes and curable by natural means.

300-1400

The teachings of Galen, Second Century Roman physician, dominated medieval medicine. In this drawing from a medieval book, he points to medicinal herbs.

1400-1600

William Harvey first demonstrated that the blood circulates continuously from the heart through arteries, into the veins and then back to the heart

PHYSIOLOGY

ANATOMY

BOTANY AND ZOOLOGY

EMBRYOLOGY

PATHOLOGY

MEDICINE

96

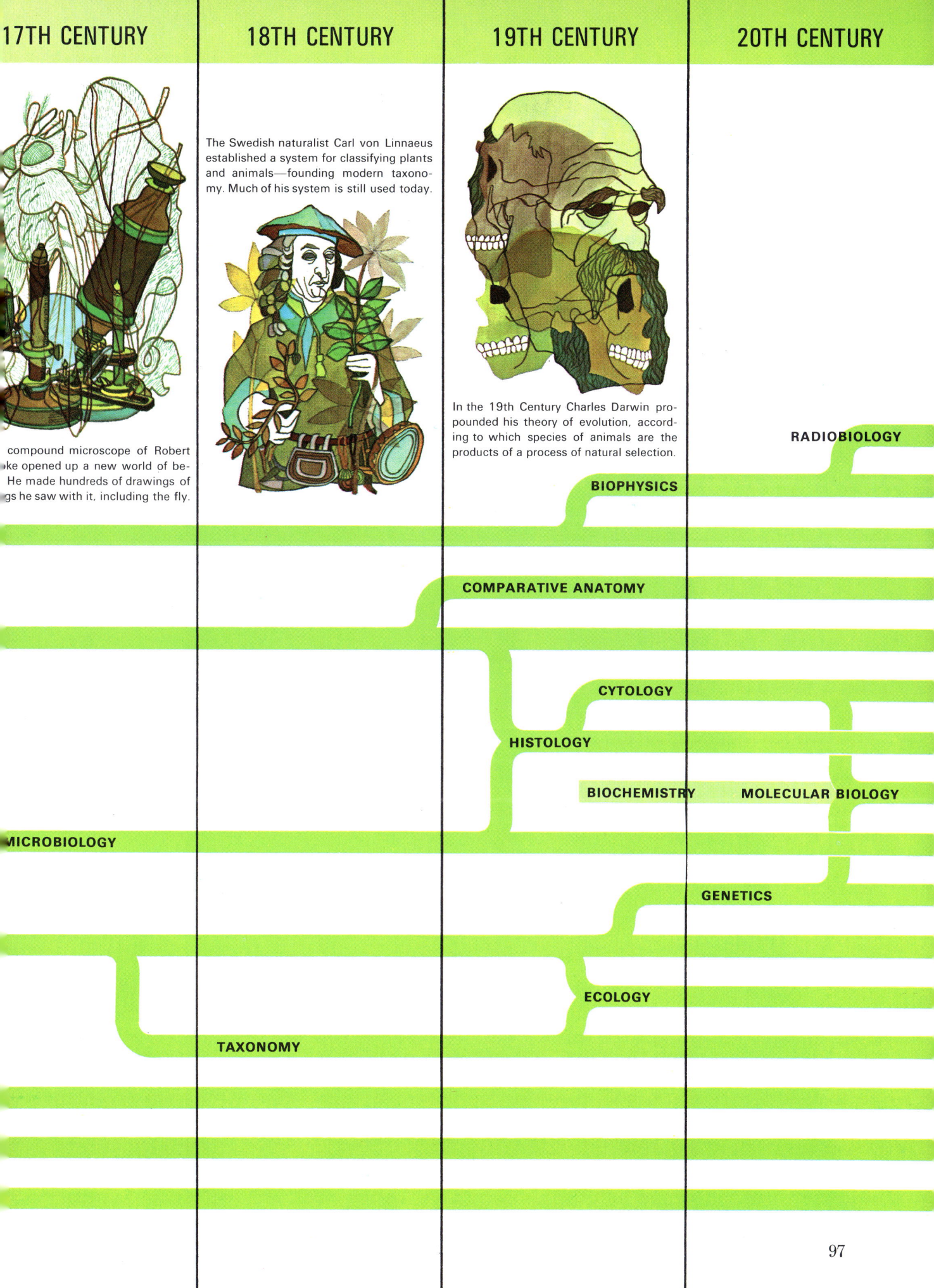

SOCIAL SCIENCES

The Study of Man and Society

Most of the branches of the social sciences, the field dealing with man and his society, are no more than 150 years old. Before that time, political thinkers, explorers, historians and philosophers—men as diverse as Machiavelli and Thomas Jefferson—had contributed to the broad mixture of knowledge and opinion about man as he was and man as he ought to be from which the social sciences sprang.

Of the four fields that fed the social sciences, only political science evolved into a science listed by the National Science Foundation. Another true science, physical anthropology, began in the 18th Century with the work of Johann Blumenbach, who first classified man into five anthropological families. Most of the social sciences became formal studies about the middle of the 19th Century, when people were dazzled by the successes of the natural sciences, and hoped that human behavior might be found to be governed by laws as simple and certain as theirs. Karl Marx, for example, saw revolutions in modern industrial nations as "inevitable" according to fixed dialectic laws.

In the same period, archeology and cultural anthropology arose to put man's past and his primitive cultures under scientific scrutiny. But of the sciences dealing with modern man, only psychology is a true experimental science and, in contrast to sociology, political science and economics, has gathered a body of knowledge that stands relatively undisputed.

2000 B.C.-300 A.D. | 300-1400 | 1400-1600

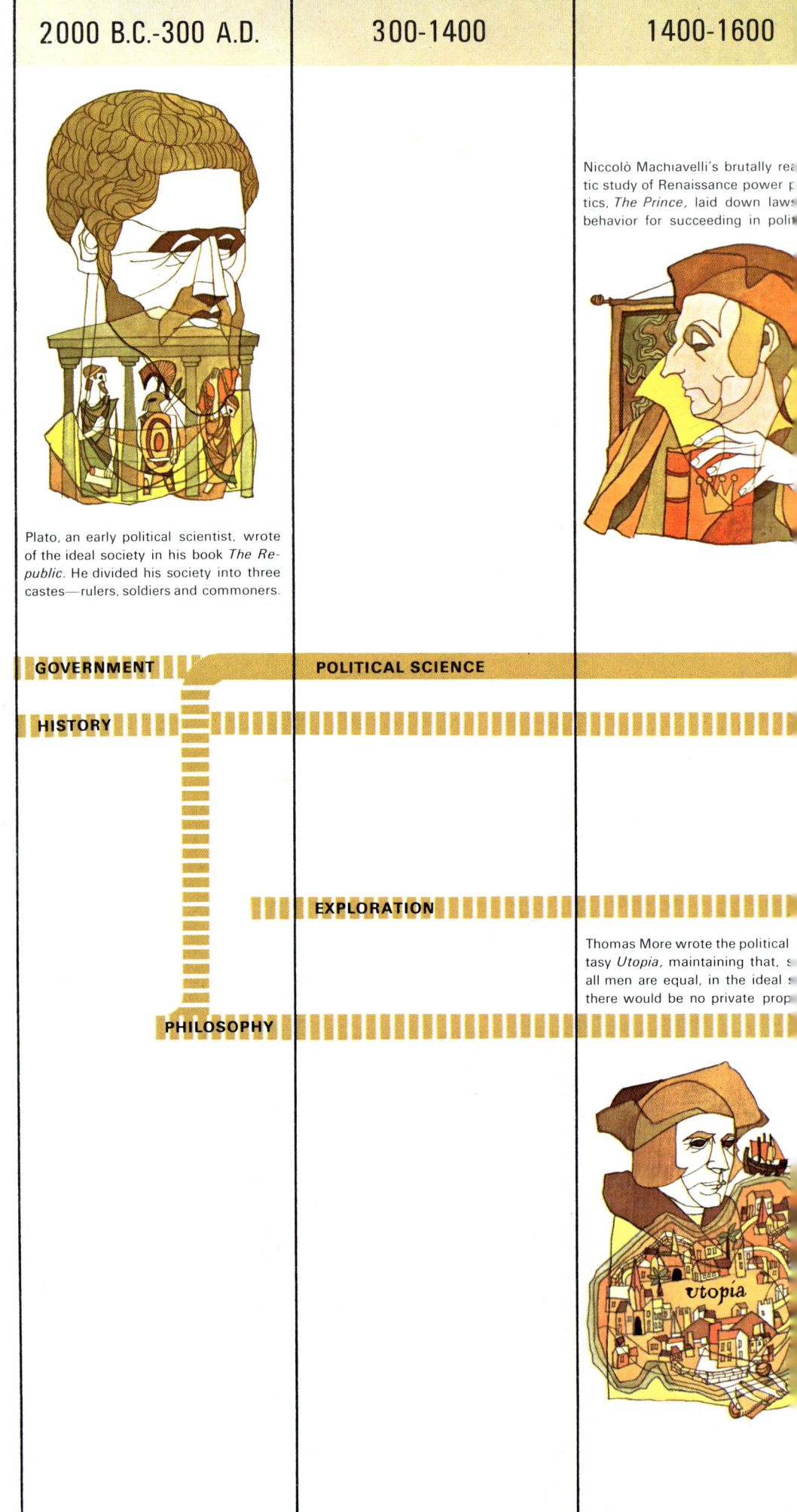

Plato, an early political scientist, wrote of the ideal society in his book *The Republic*. He divided his society into three castes—rulers, soldiers and commoners.

Niccolò Machiavelli's brutally realistic study of Renaissance power politics, *The Prince*, laid down laws behavior for succeeding in poli

Thomas More wrote the political tasy *Utopia*, maintaining that, all men are equal, in the ideal there would be no private prop

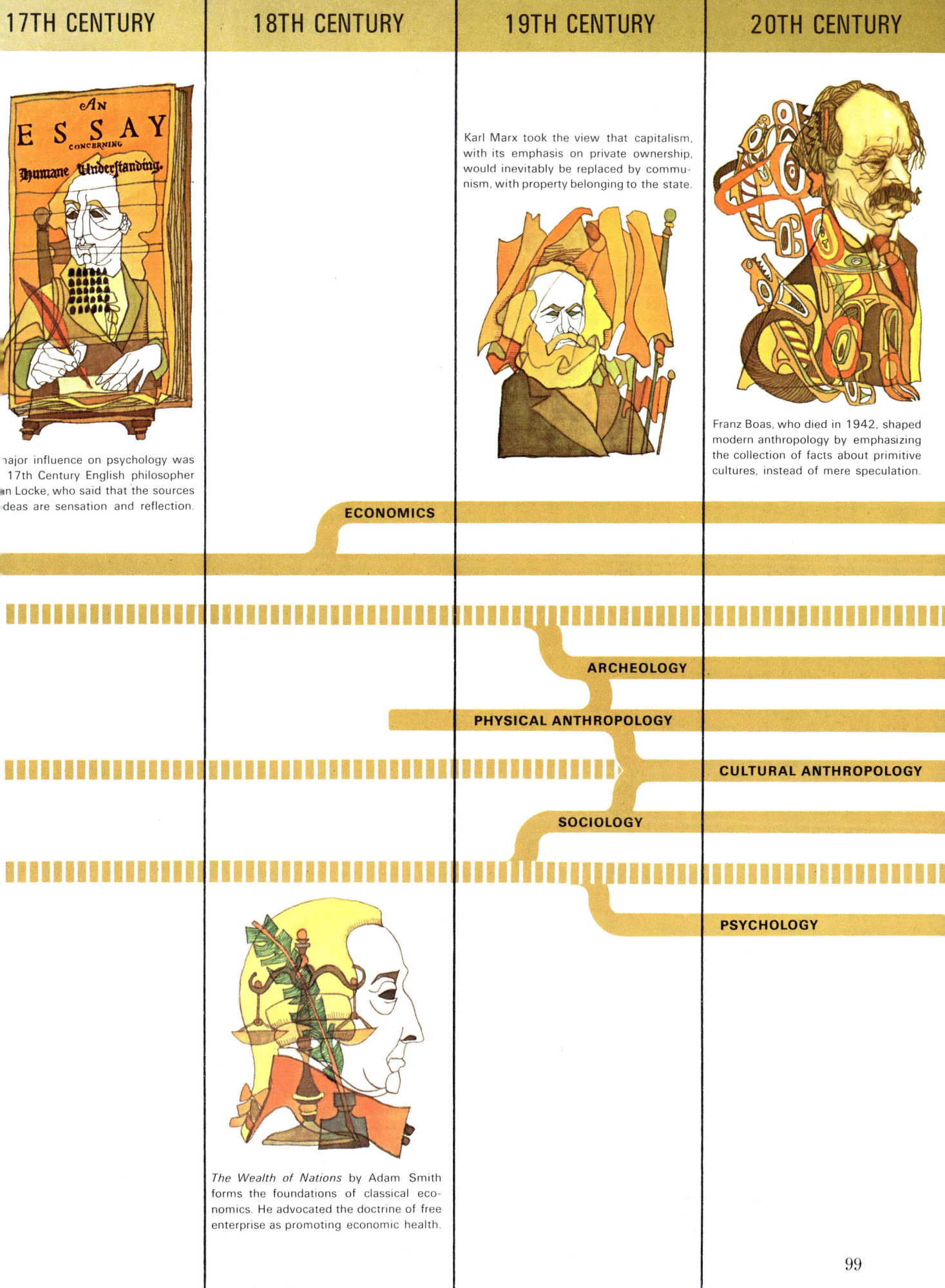

| 17TH CENTURY | 18TH CENTURY | 19TH CENTURY | 20TH CENTURY |

A major influence on psychology was 17th Century English philosopher John Locke, who said that the sources of ideas are sensation and reflection.

Karl Marx took the view that capitalism, with its emphasis on private ownership, would inevitably be replaced by communism, with property belonging to the state.

Franz Boas, who died in 1942, shaped modern anthropology by emphasizing the collection of facts about primitive cultures, instead of mere speculation.

ECONOMICS

ARCHEOLOGY

PHYSICAL ANTHROPOLOGY

CULTURAL ANTHROPOLOGY

SOCIOLOGY

PSYCHOLOGY

The Wealth of Nations by Adam Smith forms the foundations of classical economics. He advocated the doctrine of free enterprise as promoting economic health.

99

Blending the Disciplines

The many-colored chart at right illustrates the fact that the seven sciences traced on the preceding pages can and often do combine forces to explore the universe. In opposition to the proliferation of narrow specialties in each science is the phenomenal growth in recent years of the so-called "interdisciplinary" branches cutting across two or more sciences. Today every science has men who apply the knowledge developed in other sciences to branches of their own. Mathematics, of course, permeates all the sciences. Physics is also coming more and more to assume a ubiquitous presence, having such branches as biophysics, geophysics and physical chemistry.

The chart is designed to be read like a mileage chart on an automobile road map. Each science is represented by the color used earlier to trace its genealogy. Each colored rectangle is the intersection of one of the sciences at the top of the diagram with one of the sciences at the side, blending their hues and listing fields that partake of both. Where a science intersects with itself, the major branches of that science are listed.

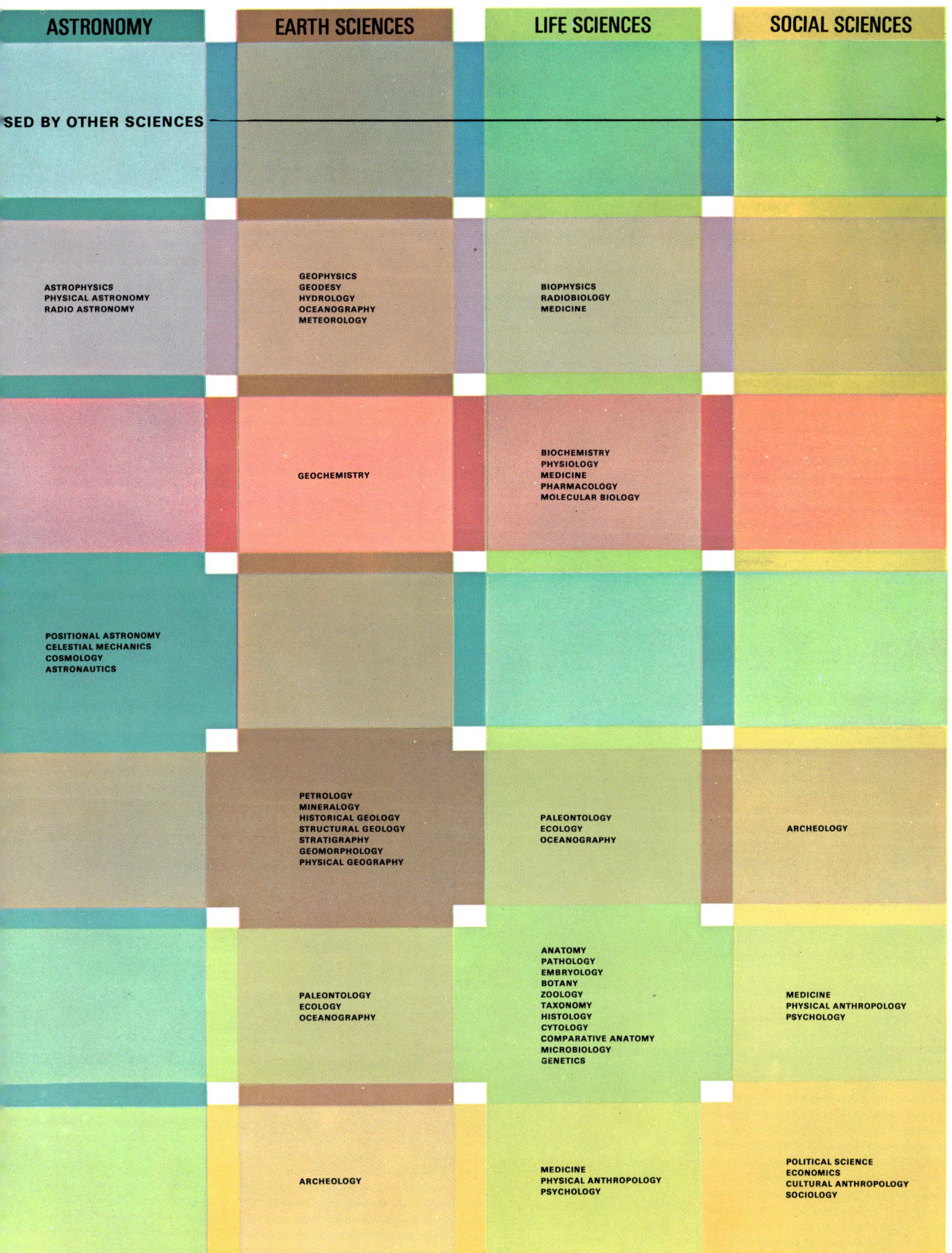

5
The Communications Gulf

OF ALL THE HARASSMENTS which plague our age, few are more typical than "crises in communications" and "information breakdowns"—those awkward silences which increasingly mar man's dialogue with man. We have trouble reaching the minds of other peoples who have been brought into elbow-rubbing proximity by the jet; trouble assimilating the headlong technological changes in our towns, homes and businesses; trouble visualizing what our neighbor does for a living when he announces himself as a product diversification specialist or a flow-meter manufacturer. Most of all, we have trouble keeping up with the basic cause of all the innovation and specialization—in other words, with science.

The obstacles to understanding science are manifold. Its most important ideas are essences of distilled essences. Its subjects, by necessity, are objective, and seldom couched in terms of "human interest." Its supporting facts are bewilderingly profuse. The last men who ventured to claim that they had all science within their ken lived about 350 years ago in the latter part of the Renaissance. Today no reputable scholar would boast that he had mastered even one sector of science, such as plasma physics or ornithology.

As if the sheer quantity and the intrinsic abstractness of science were not enough, it is made doubly difficult to grasp by the academic way in which scientists frequently express themselves. As a group they score exceedingly well on verbal aptitude tests. But like other scholars, they tend to bring to their published works much ready jargon and little simple language; much microscopic detail and little telescopic perspective; and in general, a fear of seeming opinionated, undignified or colorful. All too often, the layman attempting to read their efforts succumbs to a drowsy philistine urge to remain ignorant. Nor is his the only dilemma. Even within the scientific fold, a curtain of unintelligibility too often separates one specialist from another.

In recent years publishers, educators and scientists themselves have mounted heroic campaigns to improve communications. The gulf between scientist and adult nonscientist has been partially bridged by an increase in both the quantity and quality of science popularization. New curricula have been drawn up to improve scientific education. New techniques have been devised to ease the exchange of technical information amassed by different groups of scientific specialists.

As between scientist and nonscientist, the obstacles to understanding are most keenly felt by that articulate breed known as the humanists. Like "scientist" itself, the word "humanist" covers a multitude of meanings. Often it is used to describe a disciple of the "humanities"—language, literature, philosophy and the fine arts. More loosely, it is sometimes applied to any nonscientist. Occasionally it is equated with "humanitarian," and thus taken to imply a deeper, or at least different,

TORRENT OF WORDS
The periodical shelves of Columbia University's physics library—just one of its eight science libraries—reflect the increasing pileup of scientific communications. The collection regularly receives 470 journals in 12 languages. However burdensome the chore may be, the scientist must read extensively if he wants to stay abreast of his colleagues all over the world.

love for mankind than anyone else, scientists included, could muster.

A few years ago, the British novelist and physicist C. P. Snow suggested that science and humanism have drawn apart into two mutually uncomprehending and mistrustful cultures. The great majority of human beings, of course, are not humanists, not scientists and not even particularly cultured. But most of us, whether we realize it or not, tend to favor one side or the other. A market analyst and a carpenter, for instance, are apt to incline toward science because of their technical interests. A housewife and a politician, absorbed in civic affairs, are likely to lean toward humanism.

A legacy from Charlemagne

Although the separation between humanists and scientists has only recently become a matter of concern, its roots lie deep in the Middle Ages. In the Eighth Century A.D., the Emperor Charlemagne founded scholae throughout Western Europe. These were church schools, devoted primarily to theology. But they also offered a two-part secular program in the liberal arts—the *artes liberales,* so called because, in Roman times, their study had been permitted only to *liberi,* or free men. In the first half of the medieval curriculum, scholars took up the trivium of grammar, logic and rhetoric; in the second half, they went on to the quadrivium of arithmetic, music, geometry and astronomy. No distinction was made or intended between trivium and quadrivium; the three sciences in the latter group were regarded simply as advanced liberal arts, and were still so considered after the scholae had evolved into full-fledged universities.

During the Renaissance, however, pressures from the world outside began to create cleavages between students of the arts and sciences. Traditionally, all scholars expressed themselves in the classic elegance of Latin; now some came to see merit in the common vernacular, made increasingly popular by mass printing. Traditionally, too, scholars accepted the learning of the ancients and did not expect to add to it except by way of interpretation; yet now, scientific craftsmen such as Leonardo da Vinci, without benefit of a scholarly background, began to make notable new contributions to knowledge. These challenges to classicism met with a mixed response. The humanists of the trivium tended to cling to the old knowledge but made increasing use of contemporary idiom. The scientists of the quadrivium embraced the new knowledge and the experimental approach but tended to preserve the status quo in scholarly language, finding in the old Latinate usages the most convenient mode of communicating with colleagues in other lands. In addition, scientists have cultivated that other most international of languages, mathematics. It, too, goes back to classical times,

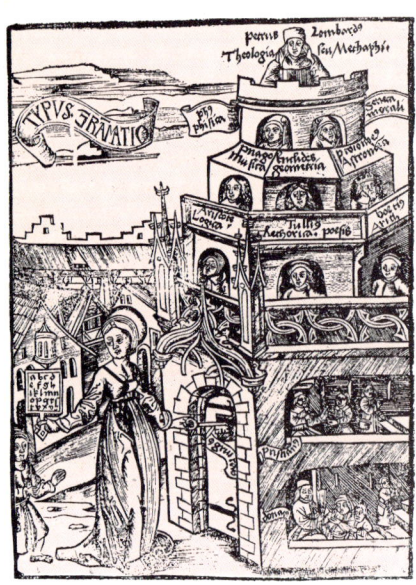

"THE HOUSE OF LEARNING" as illustrated in a 16th Century encyclopedia *(above),* showed the relationships of the various medieval academic disciplines. The two lower floors represent grammar. Looking out the windows of the third floor are, from left, Aristotle (logic), Tully (rhetoric and poetry) and Boethius (arithmetic). On the floor above are Pythagoras (music), Euclid (geometry), and Ptolemy (astronomy). Next appear Plato (philosophy) and Seneca (morality). At the top of the tower, representing the supreme studies of theology and metaphysics, is Peter Lombard, the 12th Century Italian theologian. Nicostrata, a woman who is believed to have invented the Roman alphabet, leads a student into the house through a door labeled "Harmony."

and it, too, has come to represent a formidable hurdle to the uninitiated.

In principle, all science could be translated into colloquial English. The effort would be foolish, however, for two good reasons. In the first place, the average citizen does not want or need to know all the minutiae of science, any more than he does of plumbing, versification or statecraft. In the second place, many details of science are more succinct and accurate in the original terminology—whether technical or mathematical—than they could ever be in translation. In his *Langue des Calculs* (Language of Computation), the 18th Century philosopher Etienne de Condillac tried to put two lines of algebra into simple French prose. It took him two pages, and the reading was far from easy.

By and large, what the intelligent nonscientist wants of science is some understanding of its significant ideas. He would like to know something of the general nature of the universe and its constituents, something of the major scientific theories and the technological possibilities inherent in them. This desire is often frustrated by the scientist's tendency simply to record what he has done and thought without any effort to interpret, persuade or entertain. Backed by experiment and proof, he feels that his facts and his logic will speak for themselves. The humanist, by contrast, deals in so many subjective matters that he can seldom report anything in cut-and-dried fashion. Frequently he works out his ideas by the very process of putting them into words.

The pitfalls of eloquence

From the scientist's point of view, the eloquence of the humanist often seems a clutter of threadbare frills and old wives' tales. Why say "three score and ten" when "70" will suffice? Why talk of "ill-starred lovers" now that we know that stars have no influence over our destinies? Why these persistent references to "swords and ploughshares" when men have long since done their serious fighting with guns?

The humanists, not surprisingly, are even more scathing about scientific language. Why "ecological parameters" for "conditions of life?" Why "postnatal depression" for "letdown after childbirth?" Such synonyms for simple words of ancient and honorable lineage are merely pompous, say the humanists. Further, they charge, scientific usages have left their mark on the whole of modern speech, glutting it with unlovely coinages and deadly impersonality. A businessman speaks not of "cost" but of "cost factor." An advertiser tries to tack scientific prestige onto new products by meaningless prefixes like dyna-, mega-, and hyper-. A bureaucrat notes that "uncontainerized cargo forms an attractive target to pilfering attacks" when all he means is that "loose goods are often stolen."

However cumbersome the words and phrases of science, they often have reason behind them. When the scientist says, "It was ascertained that . . ." rather than "I found that . . . ," he is using the passive construction in order to maintain an objective tone, to keep himself out of sight. When he prefers "nonadiabatic" (not incapable of passing through) to "diabatic" (through-passing), his use of the double negative represents a characteristic wariness of flat assertions. When he lifts words out of the general lexicon and gives them special meaning, he is trying not to confuse the layman but to define some fact precisely and unambiguously. To the scientist, a "radical" has no political connotations but very specifically means a group of atoms that can behave as a unit; his word "base" means not "vile" or "low-born" but a type of chemical compound.

Scientists are the first to admit that they often come off poorly in their struggles to express themselves. Yet the fact is that as a group they have tried far harder to bridge the gulf between trivium and quadrivium than have their colleagues in the humanities. During the 19th Century, while humanists and theologians for the most part merely attacked science, scientists went to considerable lengths to explain their work to the uninitiated. In the 1820s, the natural historian Amos Eaton traveled New England enlightening public audiences on biology and geology. The great English biologist, Thomas Huxley, gave regular evening lectures at the South London Workingmen's College from 1868 to 1880, expounding on Darwin's theory of evolution.

In the 20th Century, science popularization has been furthered less by men of letters like H. G. Wells than by a handful of exceptionally eloquent scientists like the astronomer Arthur Eddington, the physicist James Jeans and the biologist Julian Huxley, grandson of Thomas. Their example has encouraged the emergence of a new breed of journalist. Posted strategically between the scientist and humanist camps, they have made a profession of explaining science to the layman. The ranks of these literary middlemen have grown substantially in recent decades. In 1934 a dozen of them pioneered in founding the National Association of Science Writers; by 1964 the group's membership stood at 620.

Captive versus captivated

Until World War II, the interpreters of science had to seek out popular attention. Today their audience is more captivated than captive. The public imagination has been widely fired by such portentous developments as the atom bomb, radar, space flight and "miracle" drugs. As a result, science popularization has taken on the aspect of a genuine boom. Newspapers, for example, began to allot as much as 50 per cent more space to science after Sputnik was launched in 1957 than before. Paperbacks on scientific subjects increased from a mere 50 or so in 1949 to some 1,500 in 1959. Hardcover books on the physical and biological sciences alone tripled from 1950 to 1963.

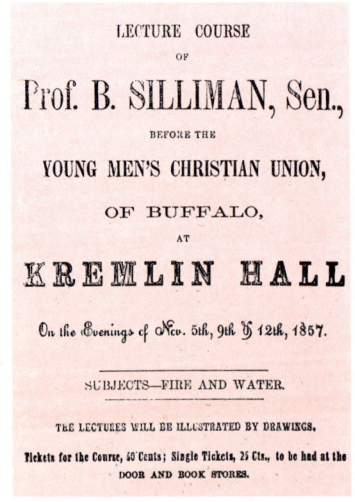

SCIENCE LECTURES, like those given by Benjamin Silliman, Yale professor of chemistry and founder of the *American Journal of Science*, packed halls in Buffalo and cities beyond in 19th Century America. Silliman, like other eminent scientists of the day, considered popular lecturing one of his obligations. He illustrated his talks with drawings, specimens and dramatic chemical experiments.

Science reporting has risen in quality as well as in quantity. The lucid science news found in magazines and in urban dailies today was relatively unknown 25 years ago. In 1940, the paleontologist George Gaylord Simpson, writing in the periodical *Science*, described a typical experience of scientists of those days in his "Case History of a Scientific News Story." Simpson had approved a routine museum press release concerning fossil discoveries in the Crazy Mountains of Montana. The release mentioned some distant primate cousins of modern man. In the various versions of the story which appeared in about 100 newspapers, Simpson found himself represented as "clambering about man's family tree," as proving Montana the cradle of mankind, and as discovering no less than 60 species, all completely fictional, of extinct giant Kodiak bear-dogs. Three years later he was still receiving letters from misinformed dog-lovers and from fundamentalists all over the country.

To this day, some scientists are leery of efforts to popularize their work. They find the demands made on them by the press burdensome. They fear that talking too freely to a journalist may stir the jibes of colleagues: "We saw your ad in the papers, Doc." They wonder whether new, untried ideas might not better be shared with the layman only after they have been applied as technology. Other scientists, however, readily sacrifice hours of leisure and privacy to the education of press and public. Their zeal is founded on the conviction that the central ideas of science are more interesting than their practical applications.

An unreportable drama

While the number of scientists who favor popularization has been increasing steadily, the popularization itself still largely lacks certain vital ingredients. The journalist seldom manages to record either the drudgery or the drama involved in the scientist's work: the frustrations of a nuclear experimenter during everyday runs on a university cyclotron, the delight of a botanist climbing his first orchid-laden tree in a tropical cloud forest, or the tension of an X-15 pilot—a fully qualified physicist—as he joy-sticks the red-hot nose of his experimental hot rod into the upper atmosphere. To the layman's great loss, these feelings are almost unreportable, except possibly in the camouflage of novels and plays.

Along with their continuing problems in getting across to the public, scientists face mounting difficulties in communicating with each other. Increasing specialization, as a result of which a herpetologist is fully intelligible only to another herpetologist, and a magnetohydrodynamicist only to another magnetohydrodynamicist, is one basic source of trouble. Another is the growth of scientific publications. It is estimated that between one and two million papers are being issued yearly in some 35,000 scientific journals throughout the world. Even in his own narrow field of

barn NUCLEAR PHYSICS: a unit of cross section. One barn is equal to 10^{-24} CM^2

cocktail party effect ACOUSTICS: the ability to focus one's attention on a specific sound in a room full of sounds

cross section NUCLEAR AND HIGH ENERGY PHYSICS: a measure of the probability for a certain reaction to occur

dope SOLID STATE PHYSICS: to add impurities (dopants) to another substance, usually solid, in a controlled manner to produce certain desired properties

moderator NUCLEAR PHYSICS: a substance such as water, deuterium, graphite or paraffin which is used to slow down neutrons to thermal energies as they collide with the atoms of the moderator

noise ELECTRONICS: the unwanted and unpredictable fluctuations which distort a received signal and hence tend to obscure the desired message. INFORMATION THEORY: items selected in a search which do not contain the information desired, or items delivered by a search through accidental code combinations

sputtering SOLID STATE PHYSICS: ejection of atoms of a substance from a surface as a result of bombardment of the surface by atomic particles

NEW MEANINGS are given to ordinary and familiar words as scientists latch onto them to label or describe new discoveries, processes, materials. The new meanings of some, as listed in the sampling above, seem farfetched in relation to the old ones: for example, "barn." Other terms, however, like "cocktail party effect," seem apt.

interest, a specialist can scarcely keep up with new developments. Yet anyone who disregards the outpouring of compressed technical information runs the risk of overlooking and duplicating costly programs of research. In 1950 a Russian mathematician, A. G. Lunts, published a paper on the application of symbolic logic to computer circuit design. During the next five years, U.S. scientists expended $200,000 of research money in rediscovering Lunts's ideas, which they could have read for the cost of translation—had they known where to look.

Against the rising storm of paper, scientists have devised a number of defenses. One is the journal of abstracts, which digests and summarizes the full-length reports in particular fields. But even these journals are too numerous for workers in hybrid specialties like biophysics and molecular biology, who must regularly take bird's-eye views of neighboring properties. Scientists have begun to think seriously of publishing journals of abstracts of abstracts—summaries of summaries.

The international hopscotch

In addition to abstracts, many scientists rely increasingly on word-of-mouth communication, often at international conferences. With bountiful travel funds more and more available from Government, industry and university, they move from continent to continent, taking occasion to consult individuals who work in their own or related corners of research. By way of these "in-groups" of global commuters, new ideas rapidly hopscotch through Pasadena, New York, Moscow, Copenhagen, Berkeley, Paris, Kyoto, Cambridge and Göttingen. Solitary researchers in less fashionable centers of science are likely to find it hard to get a hearing. One well-known case concerns a Greek electrical engineer who installed elevators in Athens as a livelihood. In 1950 he worked out a brilliant idea for improving particle accelerators through a technique known as "strong-focusing." He wrote a letter about it to the University of California's Radiation Laboratory, but it was relegated to the crank file. Several years later, after a Brookhaven team had independently discovered the strong-focusing principle, archivists disinterred the letter and the engineer, Nicholas Christofilos, was not only given credit but appointed a full-fledged physicist on the California staff.

A more rational and less risky approach to scientific communication is the automated library. The theory behind this relatively recent development is that the agility of computers may be drawn upon not only to provide almost instantaneous translation of foreign scientific works but to save time and tiring effort for a man in search of a scientific bibliography, a relevant document or a specific piece of information about work in a particular field. Computers, fed with information and instructed in indexing by a programmer-librarian, could conceivably be made to keep up with the published output of the scientific world and to maintain it in instantly accessible form. Already such machines are kept busy by NASA staffers, by researchers at the National Library of Medicine in Washington, and by scientists at a number of large industrial firms. As with most automated systems, however, this one has its drawbacks. The prospect of dipping into a console of lights and buttons, instead of a cozy shelf full of books, easily deters a researcher who merely wants to browse, to hunt for some unspecified formula that may have meaning for him outside its accepted context.

Antidotes for a flood

While some scientists see automation as their last-ditch hope for coping with the mounting flood of words, others argue that the best way to control the output is to cut it down. In many universities and industries, promotions hinge, in part, on the sheer bulk of a scholar's published papers. The pressure to "publish or perish" clutters scientific literature with items of small consequence masquerading under pretentious titles. Authors of scientific papers, say the reformers, must learn to police their own productivity, and scholarly reviewers—who traditionally do not attack another man's work unless it contains actual errors—should go further, praising genuine contributions and puncturing empty ones.

If scientists are to become more selective and lucid in their writings, and if the public is to appreciate more fully the powers and problems of science, its exhilarations and limitations, the solution must lie finally in a better basic education for both humanist and scientist. Many educators feel that by the time a student graduates from college he should, whatever his special field of interest, understand enough of every major branch of thought to be able to follow its principal advances in later years.

This educational vision is, of course, still far from a reality. But in the last two decades great strides have been made in teaching young scientists the humanities and young humanists what have been called the "humanities of science." In the 1940s, for example, the Massachusetts Institute of Technology revised its curriculum to give fledgling scientists a better grounding in the arts; Yale instituted combined majors in philosophy and physics, and Harvard set up a program of study in the philosophy, history and social effects of science, specifically designed to give future humanists a solid idea of the importance of science in civilization. Since the 1940s, scientifically oriented courses in the humanities and humanistically oriented courses in science have become almost commonplace on college campuses.

Educators see large opportunities in the new cross-fertilizing of the sciences and humanities. Pasteur as well as Flaubert can be read by a student learning French; Thomas Huxley can be perused as profitably

SCIENTIFIC JOURNALS are increasing by geometric progression. In 1750, there were about 10 on scientific subjects. At the beginning of the 19th Century, there were about 100; by 1900, there were more than 10,000. The first scientific abstract, summarizing articles in other journals, appeared in 1830, when there were 300 journals. Now there are well over 300 abstracts alone.

as Aldous Huxley in the study of English prose. Musical instrumentation and the physics of sound might well be taught together, and there seems little justification for presenting the history of science and the history of Western thought as if they were mutually exclusive. Already, on a number of campuses, combined courses in philosophy and mathematics, electronics and semantics, and anthropology and epic poetry have stirred more student interest than any of them taught separately.

The new educational approaches, the new automatic techniques for handling technical information, and the heightened efforts to interpret science in a popular way must all be pursued together if we are to prevent our society from being dominated by the narrow vision of the specialist, be he push-button warrior or self-perpetuating academician. But however intensively educators, publishers, and scientists themselves may work to make science more familiar and palatable, the layman must do his share. Understanding science will never be child's play; it will always require an expenditure of individual mental effort. Lulled by technological sales talk and coddled by technological comforts, many citizens fail to see the importance of delving into the basic scientific knowledge which made all this possible. Yet it is only by acquiring a responsible awareness of these deeper currents that they will be able to control, as well as enjoy, the social revolutions of the scientific future.

The Voice of the Scientist

In 1924, the Indian physicist Satyendrenath Bose *(opposite)* collaborated with Albert Einstein on the Bose-Einstein theory of quantum statistics—though they were half a world apart and never met until their work was completed. In their day-to-day need to exchange knowledge, scientists have developed associations, journals and even a common language in order to talk frequently with each other. But most have a growing sense that they need better communications with those outside their own fraternity. They welcome a chance to talk to the public as candidly as they do to each other. On the pages that follow, 18 scientists from the U.S. and 16 other countries are extensively quoted—some from their writings, most expressly for this book. Their voices are diverse, pungent and human as they talk about the pleasures and pains of being a scientist: their work, their relationship to other scientists and to the rest of the world.

GARLAND FOR A FELLOW WORKER
A portrait of Albert Einstein draped with a jasmine garland—an Indian gesture of reverence—looks down on Bose in his Calcutta study. Now retired, Bose was only 30 when, on an impulse, he sent some work he had done on quantum statistics to Einstein. The correspondence that followed and the publication of their joint theory brought Bose international fame among physicists.

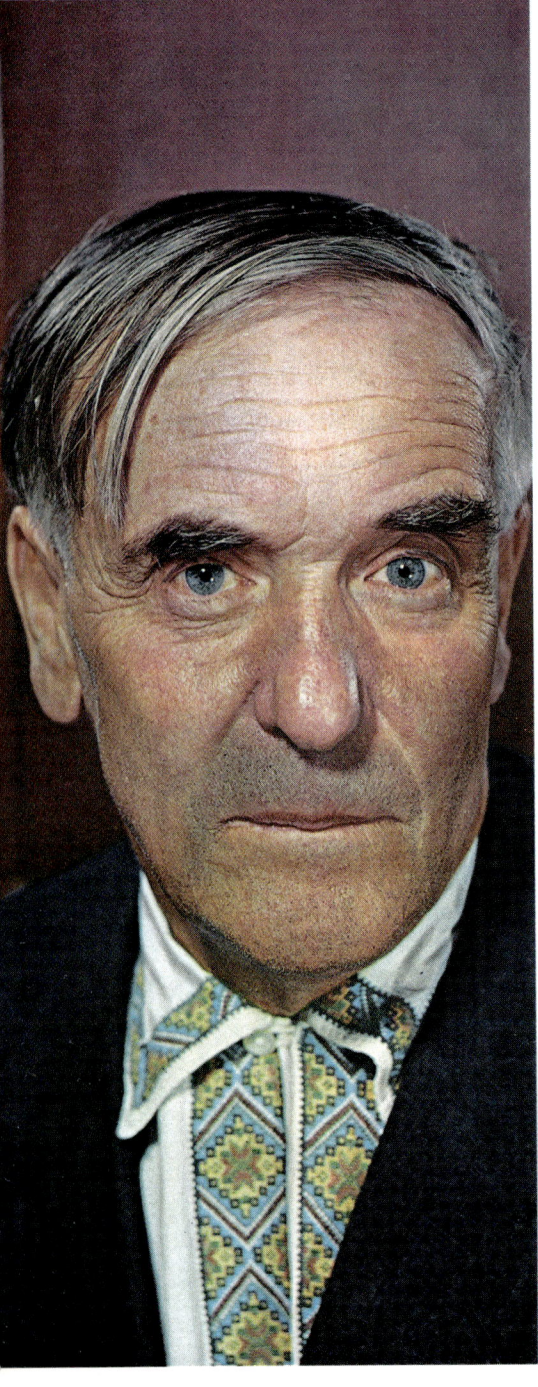

"WE ALL FEEL AT HOME"
Shown here wearing a scarf against the chill of his laboratory "cold room," Allessandro Rossi-Fanelli, head of Rome University's Institute of Biochemistry, specializes in research on the proteins myoglobin and hemoglobin. Rossi-Fanelli believes international scientific conventions "have contracted elephantiasis," but he is proud of belonging to what he calls "the great family of scientists." "We are like artists," he says, "who speak a particular language—a spiritual, fraternal, international language. When I go to Tokyo or to Moscow with biochemists, not only do we know one another, we all feel at home."

"YOU HAVE TO BE A MUSICIAN"
Peter Kapitza, a Soviet physicist famed for studies of matter at very low temperatures, was a professor at Cambridge, England, for 13 years. Returning to Russia in 1935, he was arrested in 1946 for refusing to work on nuclear weapons, then released after Stalin's death. Honored in his country once more, Kapitza criticizes Communist officials for attempting to make scientific theory fit political dogma. Science, he once told them, "is like a Stradivarius violin; this is the best violin in the world, but to play one you have to be a musician and know music. Otherwise it will sound no better than any other violin."

"OLD BORDERS...HAVE...DISSOLVED"
Manfred Eigen, of the Max Planck Institute in Göttingen, Germany, together with a colleague, Leo C. M. DeMaeyer, specializes in the physics of biological systems. "Science," says Eigen, "not only crosses international boundaries—it blurs the distinctions between once rigidly separated disciplines. A bare fifty years ago, my own field—biological physics—would have been considered an absurdity. Now the old borders between physics, chemistry, biology and medicine have almost dissolved. Findings in one area, communicated to others, sometimes set off developments and discoveries far afield."

The Bonds of a Community

As a practical matter, scientists long ago formed an international community. An Italian biochemist *(opposite, center)* notes that he feels "at home" with fellow scientists anywhere in the world. A German biophysicist *(opposite, right)* welcomes the dissolution of old barriers between disciplines. But scientists are human, and share many nonspecialized interests and concerns. An Argentinian biochemist *(below, right)* talks of the pleasures of teamwork. A Spanish physician *(below)* worries over the layman's high expectations of science. And a Russian physicist's work *(opposite, left)* can get him into political hot water.

"SCIENCE NO LONGER EXISTS"
Attended by a white-capped nursing sister, Juan Lopez-Ibor, head of Madrid University's department of psychiatry, interviews a patient. "The man in the street," Lopez-Ibor says, "once expected the scientist to interpret the universe and human life. Now he only asks the scientist to help him live, to diminish his effort and pain. The scientist is becoming more and more the technician and less and less the sage. Science no longer exists—it has been replaced by the sciences, and this dispersion of knowledge, this lack of a clear image of what is happening on earth, is one cause of today's human anguish."

"A SCIENTIST IS ALSO HUMAN"
Luis Federico Leloir—shown here in his biochemical laboratory in Buenos Aires—believes with many scientists that the day of the individual in science is over. Speaking of the 25-man staff of scientists he heads at the Institute for Biochemical Investigation, Leloir says, "We get better results this way, and it's a lot more fun. Groups can create personal problems—jealousies, frustrations. A man always needs enough freedom to pursue his own ideas and do his own thinking. But working in a team can remind us that a scientist is also a human being—and that nothing human can be indifferent to him."

"OUR TASK IS TO HELP"
Greek archeologist Nikos Yalouris hauls up a stone from excavations at the ancient site of Elis, greatest city of the Peloponnesus from the Eighth to Fourth Century B.C., where the first Olympic Games were held. Yalouris has also done pioneer work in underwater archeology off the rocky coasts of Greece and among its myriad islands. "In a country where ancient tradition is so vivid, where every day a farmer's plow digs up some piece of art from ancient times—even an occasional masterpiece—people are already well versed in archeology," says Yalouris. "Our task is to help them learn more."

"THE HUMANISTIC SIDE"
Carlos Monge, 80, now lives in lowland Lima, Peru, after three decades of medical research in the Andes Mountains. In 1927 he began making regular trips into the highest places inhabited by man—from 15,000 to 17,000 feet above sea level—studying not only altitude sickness but also the ways in which highlanders have adapted to the rigors of their life. His work is carried on by his physician son, Carlos Jr. Says Monge, "My work was based partly on my interest in the humanistic side of medicine. I am one doctor for whom the well man is more interesting and important than the sick man."

"WE . . . NEVER FEEL APART"
With his "laboratory"—the stark, active cone of Mount Asana—in the background, Takeshi Minakami stands on a lava-strewn hillside 80 miles north of Tokyo. The research of three decades enables him to predict eruptions with 90 per cent accuracy, sometimes as long as three months in advance. Minakami has stayed close to his mountain, has traveled little. But he does not feel isolated. "We volcanologists around the world never feel apart from each other. Nothing, not even politics, can interfere with our interests, for the cores of the volcanoes we study are really far too hot for anything else."

Rewards in the Backyard

"Stay-at-home" has become a scornful term for people too incurious to leave their own borders. But Takeshi Minakami, Japanese volcanologist *(above)*, Nikos Yalouris, Greek archeologist *(opposite, top)*, and Carlos Monge, Peruvian specialist in high-altitude medicine *(opposite, bottom)*, are inquisitive, restless men who have won international recognition by staying in their own backyards.

Minakami chose to study volcanoes —as small continuous earthquakes— after the catastrophic 1923 Tokyo earthquake. Monge, struck by the dearth of medical knowledge in his mountainous country about high altitudes, went to the Andes to study at firsthand the physiological effects of mountainous living on men. Yalouris turned to archeology on rediscovering the splendor of the Parthenon.

"THE SKY BELONGS TO EVERYONE"
Pol Swings, pioneer in stellar spectroscopy—the study of stars through analysis of their light waves—stands under the tube of a telescope at the teaching observatory of the University of Liége in Belgium. Swings is an authority on "peculiar stars"—stars with abnormal characteristics—but he has also used his spectroscopic skills to help Belgium's steel industry find out what happens to gases and metal alloys inside steel furnaces. "I can't very well confine my activities to my own university," Swings says. "My branch of science has no boundaries. The sky belongs to everyone, with stars to spare for all."

"THEY ALL HAVE THEIR FAILINGS"
High above the western plains of Australia, Edward Bowen climbs up to the focal cabin of the Parkes radio telescope. British-born, Bowen moved to Australia after World War II. He believes scientists should not be placed on a pedestal. "They are just like everybody else," Bowen says. "They all have their failings. Some are dedicated, some unscrupulous, some sharp as a whip, others dull as dishwater. I've known some of the great names of science, men who have done tremendous good for the world. And while I've known no scientist who's been in jail, I've known some who richly deserved to be."

Astronomers, the Happy Few

Astronomy, most venerable of the sciences, with a history dating back before written records, deals with the vastness of space, where measurements are made in units of billions. Yet it is probably the least populated community in science. Having only about 1,200 full-time astronomers, it is like a village where everyone knows everyone else, and people are always popping across town to exchange the latest news. With the universe for a subject, astronomers regard the planet earth as so small that, as one astronomer puts it, "We will hurry to the end of the globe just for a better look at a star and damn the expense."

"ON TAP, NOT ON TOP"
Vienna-born mathematician Hermann Bondi, a professor at the University of London, works harder than most at bringing his science to the public. He gives television lectures on the mathematics of astronomy. "In England," he says, "the great problem is the feeling of superiority the person educated in the humanities has for the scientist. As the saying goes, 'The expert should be on tap, not on top.' There is a tendency to steer scientists too much toward the vocational rather than the educational, and this has disqualified them from policy-making jobs in government and industry in this country."

"THE OBSERVER PLAYS A ROLE"
Vikram Sarabhai (left, above), chief of the Ahmedabad Physical Research Laboratory in India, goes over plans for a new telescope with fellow astrophysicists. Educated at Cambridge, Sarabhai specializes in cosmic rays—which most Indians have never heard of. He nonetheless feels a philosophical kinship with the largely uneducated Indian masses. "In India," he says, "there is a great deal of ignorant prejudice, which of course is alien to the scientist. But the philosophy of the ordinary Indian is not. The observer plays a large role in Indian philosophy. This is a concept the scientist can share."

117

A New Breed in Ancient Lands

Medical science is close to the earth in such countries as Iran, Egypt and Jordan, where the majority of men scratch a meager living from the soil, and where too many die young, victims of hunger and disease. The three Arab scientists shown here have contributed much to modern medicine's victories over the ancient ills of peoples in tropical climates—who make up a third of the world's population. Working doctors as well as researchers, they measure their accomplishments in terms of a child cured of anemia, a laborer made well enough to work, a mother persuaded to bring her baby for life-saving inoculations.

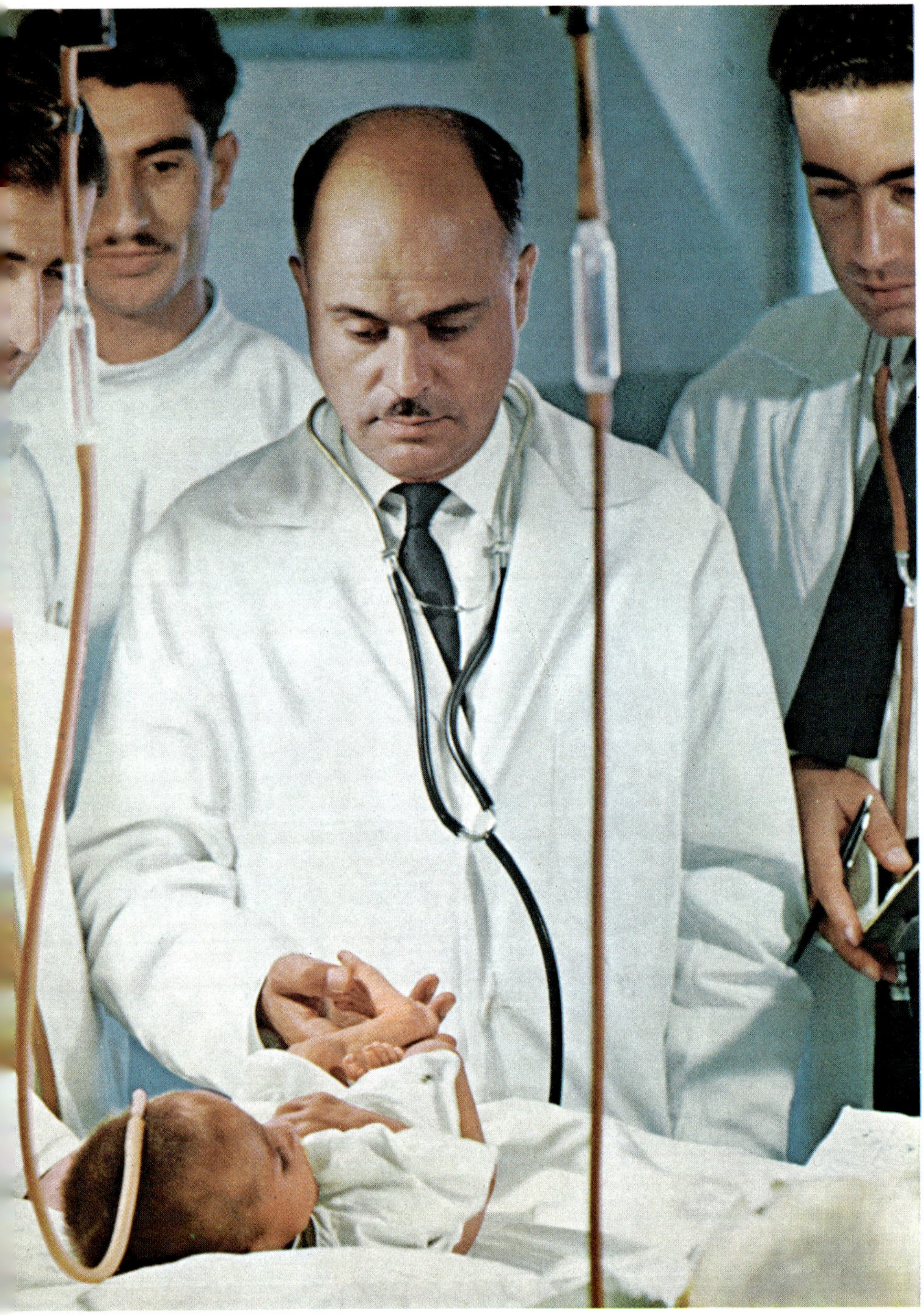

"I HAVE NO FIXED IDEAS"
At Augusta Victoria Hospital in the Jordanian sector of Jerusalem, Amin Majaj examines a tiny patient suffering from malnutrition. In 1954, without adequate laboratory facilities—he even lacked a suitable microscope—Majaj identified a widespread type of anemia that would not respond to ordinary treatment. With funds and equipment from the U.S., he is now developing a treatment for the disease. "In my field," he says, "there is not much reading to do. Most cases occur in underdeveloped countries where research facilities are poor. But it may be an advantage. I have no fixed ideas to bias my mind."

"A SCIENTIST . . . MUST . . . CRITICIZE"
Abdel Aziz Ismail—shown in his Cairo University laboratory—is a man with two contrasting specialties. Part of the time he works on eradicating bilharziasis, a disease caused by a tiny, snail-transmitted worm that has plagued Egypt since the time of the Pharaohs. He is also seeking new ways of treating atherosclerosis, the hardening of the arteries associated with the kind of advanced and affluent society Egypt is trying to become. "A scientist today," says Ismail, "must read and criticize other people's work. A knowledge gap of even a few months can mean an immense amount of futile labor."

"HARD TO BREAK THROUGH"
In the ancient Persian tongue, Chamseddine Mofidi's last name means "you are useful." As head of Tehran University's Institute of Parasitology, Tropical Medicine and Hygiene, Mofidi lives up to his name by spending much of his time in village huts like the one shown above, where every year 40 per cent of Iran's newborn children are killed by worm-caused diarrhea. "The villagers and tribesmen are wonderfully cooperative," Mofidi says. "But it is still hard to break through the ancient attitude that tells the ordinary person—hungry as he may be—that the God who gave us teeth will give us bread."

Statesmen of American Science

In the midst of the American Revolution, Benjamin Franklin, one of the first American scientists, arranged safe passage past U.S. naval guns for a British scientific expedition under Captain Cook on a voyage of "Discovery in Unknown Seas."

Since Franklin's time, American interest in international scientific cooperation has grown. Here are two examples: biophysicist Detlev Bronk, president of The Rockefeller Institute *(opposite)*, and physicist Joseph Kaplan *(left)*, of UCLA. Both have spent much of their careers leveling international barriers in science. Both men —Kaplan as chairman of the U.S. Committee for the International Geophysical Year, Bronk as president of the National Academy of Science, one of the sponsors of the IGY—can take a lot of credit for the success of the most ambitious international scientific enterprise ever undertaken.

"A TERRIBLE SHAME"
Joseph Kaplan—shown above in his study at UCLA—has recently wound up his duties as U.S. IGY chairman. He is now working on IGY's successor, 1964-1965's International Year of the Quiet Sun—a broad study in geophysics under calm solar conditions. Kaplan is encouraged by such cooperative endeavors among professionals but feels the nonscientist is still being left out. "I think it's a terrible shame," he says, "that people can live in a world where science is so important and be so ignorant about it. I call this scientific illiteracy. But many scientists don't even realize this is the problem."

"THE TRUE SPIRIT AND FAITH"
After more than 30 years of dividing his time between biophysics and administration, Detlev Bronk *(opposite)* is still trying to spend less time in his sunny office at The Rockefeller Institute and more in the laboratory. Chairman of the President's Panel on International Science, Bronk belongs to the scientific societies of six nations, from Denmark to Brazil. "The true spirit and faith of our scientific adventures seldom find their way into a scientific journal," Bronk says. "And yet they are powerful factors . . . experiences only to be gained by the free movement of scientists throughout their worldwide realm."

6
A Booming Establishment

NOT LONG AGO the word "science" called to mind a quiet backwater in which solitary men pursued their own special way of thought. Today it more often suggests vast installations, billion-dollar expenditures and a compelling voice in national policy. Science, in short, has become an establishment, one that ranks with Church or State or Business.

Big Science dates from the 1940s, and most directly from the crash-program to build the A-bomb. At the time, the physicists who took part viewed their endeavor as a patriotic chore which would end with the war. Their attitude toward bigness and organization was expressed in a lament sung at one of their first professional get-togethers after Hiroshima. The tune was an old hillbilly air, "I don't want your greenback dollars." Memories differ as to the exact words, but one version runs:

> *Take away your billion dollars; take away your tainted gold;*
> *You can keep your damn high voltage, 'cause my soul will not be sold.*
> *Take away your Army generals, 'cause their kiss is death, I'm sure.*
> *Everything I build is mine now. Every volt I make is pure.*
> *Engineering isn't physics. Is that absolutely plain?*
> *Take, oh take your billion dollars, let's be physicists again.*

Soon afterward some of the songsters returned to their modest laboratories and dusted off beloved pieces of low-voltage apparatus. Once more they set about practicing physics in the low-budget prewar tradition of "do-it-yourself," of calculations done without benefit of computer and of experiments conducted with liberal aid from baling wire and sealing wax. But the old ways were not to be recaptured. High voltages and teamwork were indispensable for many of the most exciting lines of research. Great accelerators for knocking secrets out of atomic nuclei required a massive amount of engineering and financing. National defense continued to take a tithe of time and talent from every physics faculty. The H-bomb had to be built, and the radar warning lines, and the nuclear submarines, and the intercontinental missiles, and the reconnaissance satellites. In the process, the field of physics grew out of all precedent with any period in the past. If the scientific contributors to unconventional weapons were to assemble for a songfest today, they would have to hire a Carnegie Hall.

What has happened in physics has happened to a lesser extent in most other sciences as well. The war brought home the efficacy of research with such force that scientists of every stripe have been called on increasingly to participate in the enterprises of Government and industry. No longer do they stay within ivied walls suggesting the ideas for new inventions at second-hand; they go out and invent for themselves. They join industrial concerns or even found businesses of their own. They are directly responsible for our latest generation of rockets,

CITADEL OF SCIENCE
A Mount Olympus among scientific fraternities, Britain's Royal Society *(left)* has flourished for more than three centuries. Only 29 new members are admitted yearly. The Society's publications, now read the world over, were launched in 1665 with a monthly journal which included, among other items, a report on "The New American Whalefishing about the Bermudas."

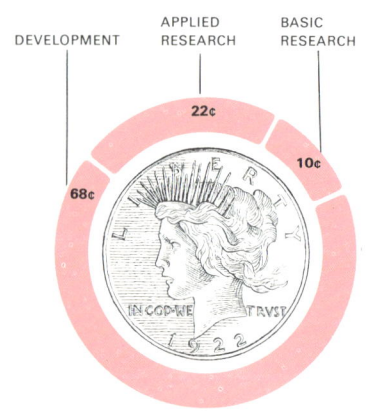

THE SCIENTIFIC DOLLAR spent in the U.S. during 1961-1962 was allocated this way. Ten cents went to basic research—which grapples with fundamental theoretical questions. Applied research, which looks for practical uses for the findings of basic research, received 22 cents of every dollar. Most of the money, 68 cents, went to development—which manufactures and tests new products.

drugs, plastics, ceramics, alloys, electronic gadgets and automatic factories. Many a sector of industry would face slowdown or shutdown without its staff of scientific Ph.D.'s. So heavily have scientists become involved in the world of affairs that as a group they find proportionately less time for basic research; U.S. expenditures for this purpose have dropped about 40 per cent—from roughly 17 cents out of every dollar spent for all phases of science in 1940 to 10 cents out of the dollar in 1962.

The main change that has overtaken science, however, is sheer, prodigious growth on every front: basic or applied, theoretical or experimental. The annual outlay for U.S. scientific and technological activity rose from about $350 million in 1940 to over $16 billion in 1963—a 47-fold increase. Over the same period the number of scientists tripled. By the end of 1963, we had 255,000 physical scientists and mathematicians, 160,000 life scientists and 85,000 social scientists. We also had 250,000 high school science teachers and 1,000,000 technicians and lab assistants. Counting in 935,000 engineers as well, science has become entrenched in our midst to a total of 2,685,000 workers, representing 3.6 per cent of the entire labor force—as compared with only 1.5 per cent in 1940.

The saga of a platoon

Worldwide, the scientific population explosion has dwarfed the general population explosion as atomic fission dwarfs a detonation of TNT. Indeed, science itself has largely triggered both bangs. Since the Renaissance, the world's population, over-all, has doubled and redoubled. Over approximately the same period, the scientific community, by contrast, has multiplied a hundredfold each century: from a platoon counted in individuals in 1665, to a regiment counted in hundreds in 1765, to a division counted in ten thousands in 1865, to an army estimated at six million today. This phenomenal rate of proliferation has meant that the number of scientists living at any one time has constituted 90 per cent of the total number of scientists who have ever lived. Thus we have nine times as many scientists now as in all previous eras put together. As a corollary, it can be reasonably estimated that nine tenths of current scientific knowledge was yet to be created when our present elder statesmen of science graduated from college in the 1920s.

The growth of science—not only in men, but in discoveries, devices, journals, papers and so on—has inevitably led to administrative and financial complexities. On the whole, these have been dealt with in impromptu fashion, one by one, as they have arisen. As a result, the U.S. scientific establishment today is a sprawling tangle of societies, bureaus, departments, boards, institutes, and research centers presided over by interlocking directorates of committees and subcommittees. This is hardly a tidy setup, but there are reasons for it.

In science, growth depends basically upon the initiative of creative individuals who open up new fields of research. Such men will find funds and facilities with which to pursue their interests by hook or by crook, if necessary; when there is no place for them in an existing structure, they make one. The habit is deep-rooted. During the 1830s, for example, U.S. astronomers sought to have Congress set up a national observatory. Congress turned down the proposal, but the astronomers were not to be deterred. The Navy shortly thereafter established a "Depot of Charts and Instruments" to house, among other equipment, the chronometers which helped fix the positions of ships at sea. Small-scale astronomical observations were required to keep the chronometers accurate. Astronomers on this project gradually convinced their chiefs of the practical value of their craft, and in 1841, the head of the Depot persuaded Congress to appropriate money for a more adequate building. This turned out to have an elaborate dome which could be opened mechanically to the sky and was just right for telescopes. Once it was an accomplished fact, the building became the home of the U.S. Naval Observatory.

Today scientists marshal their forces in four main types of organization: societies that keep them in touch with one another; associations through which they maintain liaison with the Government and with the public at large; great laboratories and institutes in which they come together to do research; and universities in which they come together to teach. Aggregations of scientists are, of course, not new. For example, the societies—which now number more than 1,500 in the U.S. alone—evolved as far back as the 16th Century out of informal groups of "natural philosophers" who enjoyed sharing a meal, a few drinks and shop talk. One of the oldest of these groups in continuous existence is the Royal Society in England, chartered by Charles II in 1662 and built around a coterie of such distinguished figures as the chemist Robert Boyle, the diarist Samuel Pepys, the architect Christopher Wren and the physicist Robert Hooke. Isaac Newton served from 1703 to 1727 as the Society's 13th president. (In his honor, it has always sat on Thursdays since 1703, instead of Wednesdays as before, for Newton was master of the British mint, and Wednesday was the day the mint paid out coin.)

A boost from Ben Franklin

Among the members of the Society in the New World was Benjamin Franklin. In 1727 he decided that the colonies should have their own scientific association, and he founded a secret fraternity of intelligentsia initially called the "Junto" and later renamed the American Philosophical Society. This prestigious group was followed in 1780 by the almost equally influential American Academy of Arts and Sciences and in 1848 by the American Association for the Advancement of Science.

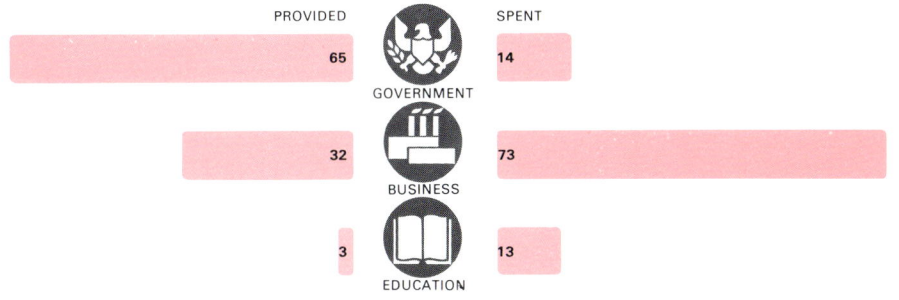

SOURCES OF FUNDS for scientific research and development (R&D) are the federal Government; business and industry; and educational institutions and other nonprofit organizations, which together provided $14.7 billion in 1961-1962. The bar graph on the far left shows the percentage provided by each source. Much of the Government's contribution went to the other two groups. The graph to the near left shows how much each group spent directly on R&D—with industry and education spending more than they provided.

The "Triple-A-S," as it is informally known, and its 304 affiliated societies now comprise the world's largest scientific organization, with a combined membership of more than two million.

In addition to being organized nationally, scientists are also frequently organized by state, city and professional specialty. Most sciences have their own fraternities, such as the 97,000-member American Chemical Society. These groups, in turn, are often affiliated with world organizations such as the International Union of Pure and Applied Chemistry.

Brainpower at work

From the beginning, many of the learned societies have advised their governments on technological matters. The French counterpart of the British Royal Society, the Académie des Sciences, was consulted by the Bourbon kings on a wide variety of problems from cannon-casting to the design of the fountains at Versailles. Later, in Republican times, its members accompanied Napoleon en masse on his military campaign in Egypt, in order to study ancient civilization and human origins.

The French example intrigued the American founding fathers, several of whom, including Franklin, Jefferson and Madison, were practicing part-time scientists. As a result, Congress was empowered by Section 8 of the Constitution to "promote the Progress of Science" by protecting inventors and authors. However, it proved leery of interpreting its franchise broadly. To get Congressional approval of the Lewis and Clark Expedition, President Jefferson had to call it a purely commercial venture, even though he had conceived it partly as a scientific one and had so represented it to the French and Spanish Ambassadors.

The need for such stratagems disappeared as awareness grew of the role science could play in various public endeavors. Gradually the Government itself began to become permanently involved in scientific research through the activities of the Weather Bureau, the Patent Office, the Coast and Geodetic Survey, and the Army Corps of Engineers. In 1863 Congress chartered the National Academy of Sciences, giving it the responsibility to "investigate, experiment and report" on scientific subjects "whenever called upon by any department of the Government." This institution is now the ranking organization of U.S. science; the number of new members admitted yearly is limited to 35 Americans and four foreigners—all chosen for their achievements in basic research.

Since 1863, in times of crisis, the Government has regularly mobilized scientists in teams of technical advisers: the National Research Council during World War I, the Office of Scientific Research and Development during World War II and the Science Advisory Committee during the Korean conflict. Since Sputnik was launched, this committee has occupied offices in the White House, and its chairman holds the top advisory post of Special Assistant to the President for Science and Technology.

The Government also maintains long-term programs to support basic research. Agencies that disburse money for this purpose include the Office of Naval Research, the Air Force Office of Scientific Research and the National Science Foundation. Set up in 1950, the Foundation sponsors about 10 per cent of the nation's basic research, spending some $150 million annually for this purpose. To help it make its allocations, panels of scientists, serving without charge, read countless applications for grants and recommend those that seem worthy.

What seems worthy to a panel of specialists does not always seem worthy to a legislator. In 1962, for instance, Senator Harry Byrd raised a storm when he discovered, in the budget of the National Institute of Mental Health, an item of $1,201,925 to be spent over a period of five years for an investigation into the love of baby monkeys for their mothers. The appropriation was approved only after psychologists explained that monkeys suckled by artificial mothers made of terry cloth develop simian neuroses which help clinicians in dealing with the problems of human orphans.

Although scientists have a high batting average in their innings with Congress, the battle is never ended and is not likely to be. Some legislators tend to view scientists as an incomprehensible and occasionally condescending breed, and some scientists, in turn, say that legislators are illiterate about science and far too impatient in their expectations of basic research.

The strongholds of research

Most of the research of science today is carried on at great centers that are exclusively a 20th Century creation. These laboratories and institutes, variously maintained by universities, by philanthropic endowments, by Government or by industry, function within the U.S. scientific establishment as powerful autonomous states. Among them are the National Institutes of Health in Bethesda, Maryland; the Institute for Advanced Study in Princeton, New Jersey; the Mount Palomar-Mount Wilson complex of telescopes outside Los Angeles; the Center for Advanced Study in the Behavioral Sciences at Stanford; the Atomic Energy Commission installations at Brookhaven, Oak Ridge and Los Alamos; and the grassy retreats set apart for research staffers by major corporations in such fields as electronics and chemicals.

The facilities provided at such centers are a far cry from the laboratories of the 19th Century. These, for the most part, contained many small cubicles and many different pieces of apparatus, devised as need arose by scientists engaged in individual experiments. Many of the laboratories of today, by contrast, are built around clusters of permanent multi-

PIONEERING WOMEN, chemistry students at the Massachusetts Institute of Technology in 1869, are shown in a contemporary newspaper sketch with their instructors. M.I.T., which enrolled women at its founding in 1859, was one of the first institutions to require science students to duplicate famous laboratory experiments.

purpose equipment or around mammoth single instruments such as the 33-billion-electron-volt synchrotron at Brookhaven. Installations of this size require teams of workers for their operation and, of course, healthy budgets. They exist not only in the realm of physics but increasingly in other fields of science as well. Cytologists working behind glass walls tend bottles of living cells in capacious tissue-culture labs kept sterile by elaborate air locks and filtration systems. For biologists studying the life processes of plants and animals, the University of Wisconsin is planning a $4.8 million "biotron," a kind of super-greenhouse with such refinements as "desert rooms," "dew rooms," "chemical atmosphere rooms" and "wind rooms." Some of the elaborate new workshops of science are even mobile. Recently built for the Woods Hole Oceanographic Institute in Massachusetts, the 210-foot ship *Atlantis* propels four spacious biological and geological laboratories through the earth's seas. Her cranes and winches lower packets of instruments and retrieve doubloons of information from as deep as seven miles down. Her aquaria make fish from all climes and fathoms feel at home. Her bow bulges underwater with a six-port observation cabin from which powerful spotlights can be played on creatures of the murky waters ahead.

The cost of such facilities grows ever larger. Just how expensive research equipment has become may be judged by one example from the ancient science of astronomy. The little "optick glass" which Galileo turned on the heavens in 1609 cost the equivalent of a few dollars. The most spectacular 19th Century telescope, the 72-inch "Leviathan of Parsontown" in Ireland, was built privately by the Earl of Rosse at a cost of approximately $150,000. The 100-inch telescope at Mount Wilson, completed in 1917, cost $600,000. The 200-inch telescope at Mount Palomar, completed in 1947, cost $6.5 million. More recently, an attempt by the U.S. Navy to build a 600-foot steerable radio telescope—cousin to the optical telescope—was abandoned after $42 million had already been expended on steel and concrete and access roads; its completion, at an estimated total cost of $200 million, was made pointless by advances in satellite technology.

An affair of economics

Many current installations can only be financed by a combination of funds: governmental, industrial, philanthropic and academic. This harsh fact of economics has in itself further complicated the scientific establishment. In some instances a number of universities may share such facilities, and so band together formally to administer them. This practice has given rise, for example, to Associated Universities, a corporation which represents nine universities in the East and administers both Brookhaven and the National Radio Astronomy Observatory at Green

Bank, West Virginia, under contract to the Atomic Energy Commission and National Science Foundation respectively. In other instances, one university alone will run an AEC-owned installation; the University of Chicago, for example, operates the Argonne National Laboratories, while the University of California operates the Lawrence Radiation Laboratory outside San Francisco. The Mount Wilson and Mount Palomar Observatories are jointly financed and operated by the Carnegie Institution and the California Institute of Technology. In yet another arrangement, Caltech operates Jet Propulsion Laboratories for the National Aeronautics and Space Administration, while the University of Arizona owns its Lunar and Planetary Laboratory but derives all the financial support for it from NASA. And so the tangle goes—and grows.

When the individual scientist—still the mainspring of modern science—needs money to pursue his investigations, obviously he must know his way around the organizational maze. He may apply to Government agencies which disburse research funds. Or he may draw on the funds of his own university. Or he may apply to a private philanthropy like the Rockefeller Foundation or the Ford Foundation. Or he may offer his services, temporarily, to an industrial-research lab maintained by companies like Bell Telephone and International Business Machines. He pays a price for his support: first, he must write up his "project proposal," then he must file monthly, quarterly or semi-annual progress reports—a chore that seldom appeals to the unregimented.

Boom and aftermath

The stupendous boom in science in recent decades has raised the inevitable question: how much longer can it go on? The answer, by general agreement, is that the growth will soon begin to level off. The prophets base their judgment on a number of statistical signposts. Educators estimate that the high-IQ sector of the population that can be profitably trained as scientists will have been fully tapped in another generation. Economists calculate that if the amount of money spent annually on research and development continued to rise at the present rate, the sum would begin to exceed the entire projected gross national product in the year 2000. Other projections of present trends are no less staggering to contemplate. If the output of scientific papers were to follow the current curve, the heft of one periodical alone—*The Physical Review*—would begin to exceed the weight of the earth itself in the mid-21st Century. If the scientific population were to continue its present rise, scientists would begin to outnumber people in the 22nd Century. Not even the most sophisticated mathematicians see any plausible way to imagine such situations, and so in the next 20 to 30 years scientific growth in terms of both manpower and money is expected to slacken and fall into regular step with the growth of the whole human enterprise.

The shortage of scientific talent is already in evidence. Recently, for instance, the American Institute of Physics predicted that the number of physicists now in training will fall 20,000 short of the jobs available for them in 1970—this despite the fact that median salaries for graduate physicists in industry rose from an average of $6,100 a year in 1951 to $11,000 in 1962. Such shortages can probably be offset in the future only by an increased pooling of international scientific resources.

Curiously, many of the most thoughtful scientists are not unduly concerned over the prospect of a slowdown in scientific growth. Indeed, they would welcome a less spectacular and more orderly kind of advance. They regard many of the developments of recent decades as abuses. A scarcity of scientists, they think, would compel a higher evaluation of basic research, save precious talent now spent on weaponry, eliminate waste and duplication in industrial efforts, and even help sort out and simplify the labyrinths of scientific organization.

These men believe, in sum, that the scientific explosion has been largely technological, and that it has not been accompanied by a comparable increase in basic knowledge. They say that such knowledge, like good wine, needs care and aging, and that this process can best be furthered if science regains its old scholarly idealism and detachment.

California, "the Science State"

California's claim to the title of "the science state" is based on a scientific establishment of awesome size and wealth. Its aerospace industry, much of it clustered in the Los Angeles area *(opposite)*, boasts more than a third of the nation's aviation and space workers. In the San Francisco area alone, electronics is an $800-million-a-year business. But at the heart of California's science empire are its colleges. While industry and the military contribute cash to California's science empire, its colleges are contributing trained manpower. The California Institute of Technology near Los Angeles is a renowned pure-science university. The nine campuses of the University of California are a steady source of science graduates. Thanks chiefly to its educational institutions, California recently had 12 per cent of all U.S. scientists, 36 per cent of the world's Nobel Prize winners in science—and the manpower for continued scientific growth.

WHERE THE SCIENTISTS ARE
Besides huge concentrations of diversified scientific activity around Los Angeles and San Francisco, California has numerous special scientific establishments: e.g., rocket-engine manufacture in Sacramento, satellite-tracking at Camp Irwin, aerospace research at Edwards and Santa Barbara, missile-testing at Point Arguello, an observatory at Palomar, ocean studies at San Diego.

LEAKY OLD FAITHFUL

By 1932, only two years after his first four-inch cyclotron, Lawrence had graduated to this one, which had a 27-inch vacuum chamber. Funds for it were difficult to raise; its magnet was donated by the Federal Telegraph Company; its vacuum sealer was a leaky mixture of wax and rosin that required endless hours to repair. Nevertheless, much experimental work was done with it, including the bombardment of a live mouse with neutrons—marking the start of the lab's long interest in medical physics. The cyclotron remained in harness until 1946.

PIONEERS IN NUCLEAR RESEARCH
This 1938 portrait shows staff members of the radiation laboratory at Berkeley, where cyclotrons were developed, posing inside the newly arrived magnet for a 60-inch cyclotron—the lab's biggest cyclotron to date. Lab director Ernest Lawrence is seated front row, center.

FIRST OF THE CYCLOTRONS
The ancestor of Berkeley's cyclotrons was this 4.5-inch brass vacuum chamber, sealed with red wax, built by Lawrence and a colleague in 1930. The chamber was hooked to an oscillator and placed between the poles of a magnet. The whole apparatus operated on house current.

The Atom Smashers of Berkeley

California's participation in the field of atomic research was assured in the 1930s when the youngest professor on the Berkeley campus of the University of California, physicist Ernest O. Lawrence, developed, with help from a graduate student, the first effective cyclotron *(above)*—a device for accelerating atomic particles in a circular vacuum chamber. From this beginning grew the Lawrence Radiation Laboratory, which was founded in 1936 and has been a major influence in shaping the atomic age ever since.

The pictures on these pages show the lab's growth in the '30s, when Lawrence and his staff wangled munificent $1,000 grants to build bigger cyclotrons, used beeswax as a sealer and five-gallon cans filled with water as radiation shields. The following pages show its postwar growth. The lab's first decade was crowned when Lawrence won a Nobel Prize in 1939.

GIANT TOOL FOR NEW RESEARCH

The latest in a long line of cyclotrons at the Lawrence Radiation Laboratory is the bevatron, so called because it generates power measured in billions of electron volts. When asked what secrets it might reveal, Lawrence said, "If we knew that, we wouldn't have built it." Following its completion in 1954, it uncovered many subatomic particles and radioactive isotopes.

The Postwar Face of Science

In 1939, when he won the Nobel Prize, Ernest Lawrence's laboratory staff numbered 46. By 1940 Lawrence and his staff had plunged into bomb research. They helped organize the Manhattan Project, and produced in the lab's 60-inch cyclotron the material transuranium, later used to explode the first A-bombs.

In 1946 the Atomic Energy Commission was formed. The radiation laboratory, with a staff now grown to some 9,000, has been wholly subsidized by the AEC ever since. Two atom smashers have gone up on the Berkeley campus *(below)*, and a weapons research branch has been established at Livermore, California.

In 1958 Lawrence became ill while attending test-ban treaty negotiations in Geneva. He hurried home to die. But he left behind a huge scientific establishment in California that still helps shape the postwar world.

A TEAM OF BIG WHEELS
The round buildings overlooking the Berkeley campus in the photograph above shelter the Lawrence Radiation Laboratory's two postwar atom-smashing cyclotrons. The smaller building on the hilltop at left houses a 184-inch synchrocyclotron, built in 1946. The big bevatron, which occupies the larger structure in the center, generates 6.2 billion electron volts.

Partners in Science at Palo Alto

Palo Alto owes its eminence as an electronics center to a remarkable collaboration between industry and education. Several radio and telephone companies were already established in Palo Alto in 1912, when Lee De Forest, inventor of the triode radio tube, joined two associates there. Their experiments with his tube launched the electronics age in the front yard of Stanford University. Stanford plunged into the field in 1924, setting up a communications laboratory, with Dr. Frederick E. Terman as director. Terman became known as the father of electronics research at Stanford—and, affectionately, as the father of Stanford's young researchers.

Stanford's scientists nourished Palo Alto's growth—and science's. In 1937 a doctoral student in physics, Russell Varian, developed the klystron tube, used in today's radar and television. Other momentous inventions by Stanford scientists include the electron microscope, the mercury vapor-tube light and the laser beam.

The joint efforts of Stanford and industry broadened after World War II. In 1946 the university created the nonprofit Stanford Research Institute, today the world's largest applied-research organization. Stanford Industrial Park, perhaps the first major real-estate project developed by a university for industry, was opened in 1950 with the signing of a contract by graduate Russell Varian's electronics firm, Varian Associates. The Park quickly attracted other high-technology companies, most of which had scientists from Stanford in key positions *(opposite)*. By 1964 some 400 acres were occupied by 42 firms with more than 10,000 scientific workers.

CLASSIC CAMPUS FOR NEW IDEAS
Stanford's serene campus and Romanesque architecture belie the university's vigorous trailblazing in science and technology. These buildings, part of the original campus, date back to the 1890s. The 285-foot tower *(background)* was added in 1941 and named for former President Herbert Hoover, who was a geology major in Stanford's first graduating class in 1895.

BIOLOGIST-INVENTOR
Dr. Joshua Lederberg examines a model of his "Multivator," designed to be rocketed to Mars to determine if life exists there. A Nobel laureate, he is head of Stanford's genetics department and director of the Lieutenant Joseph P. Kennedy Jr. Laboratory for Molecular Medicine.

PHYSICIST-INVENTOR
Dr. Arthur Shawlow, co-inventor of the laser beam, displays its power in a classroom demonstration with a converted toy ray gun. The beam's intense red light passes harmlessly through the colorless outer balloon. But the blue inner balloon absorbed the light and burst.

FROM THE STANFORD FACULTY
Stanley F. Kaisel, formerly a Stanford lecturer *(above),* is president and founder of Microwave Electronics Corporation, which designs and makes instruments for space communications.

FROM THE RESEARCH INSTITUTE
A member of the Stanford Research Institute for seven years, John V. N. Granger of Granger Associates uses his background in manufacturing electronics components and systems.

FROM THE CLASS OF 1934
William R. Hewlett and David Packard, Stanford '34, are cofounders of the Hewlett-Packard Company, now the largest producer of electronic measuring devices. Their first triumph: Walt Disney bought their audio oscillators for creating sound effects in his movie *Fantasia.*

FROM THE GRADUATE SCHOOL
Dean Watkins, a Stanford Ph.D. in physics, heads the Watkins-Johnson Company, whose Industrial Park plant manufactures electronics equipment for space-communications systems.

UCLA: HUGE AND DIVERSIFIED
UCLA's campus, seen in this aerial view looking toward the southwest, stretches over 411 acres. A new science-oriented campus is now being built to the north, out of sight in the picture. UCLA currently has 24,000 students, of whom about 30 per cent are science majors.

A NOBEL LAUREATE IN CLASS
Professor Willard Libby, 1960 Nobel Prize-winning chemist, talks thoughtfully with UCLA students in a postgraduate seminar. A dedicated teacher, Libby once said that developing a few creative students "means as much to me as discovering an important scientific principle."

Two Ways of Teaching Science

Around Los Angeles, leadership in science education is shared by two very different institutions. One, the University of California's Los Angeles campus *(opposite)*, is a huge general-studies complex whose science departments rose to prominence only recently. UCLA's neighbor, the California Institute of Technology *(right)*, is small, specialized and fiercely devoted to pure science research.

Caltech's unique reputation rests on its encouragement of originality and of the search for knowledge. Its 525 faculty members, more than half of them nonteaching researchers, range freely through all the sciences, crossing departmental lines that are held sacrosanct at many colleges. Caltech has not ignored applied science: it served as midwife to the local aircraft industry and now runs the Government's Jet Propulsion Laboratory. But it is as a community of eager purists that Caltech pursues its ideal of "contributing to the advancement of science and thus to the intellectual and material welfare of mankind."

CALTECH: SMALL AND SELECTIVE
White-domed Beckman Auditorium rises in the foreground in this photograph of the Caltech campus. The tough little university of 1,500 students limits each freshman class to about 200.

A DOCTORAL GROUP AT LEISURE
At Caltech's pool, two Ph.D. candidates chat with Dr. Robert P. Dilworth (in water, *center*), Professor of Mathematics, and Dr. Foster Strong *(seated right)*, physics professor and Dean of Freshmen. It has been said of Caltech, reflecting its convivial atmosphere, "Undergraduates are treated like graduate students, and the graduate students are treated like colleagues."

New Plants for Tomorrow's Science

Research facilities, the touchstone of tomorrow's applied science, are today costly beyond the imagination of yesterday's planners. But growth feeds on growth. Because California possesses a vast science establishment, more money is spent there than in any other state for new research facilities. In 1963 these funds included $4.6 billion from the Government, representing 38 per cent of federal research and development contracts.

To further high-energy physics, a prodigious atom smasher is now under construction at Stanford University *(left, top)*. This linear accelerator is designed to generate seven times as much energy (45 billion electron volts) as Berkeley's circular bevatron; and it will cost almost 12 times as much to complete—$114 million.

An even greater installation is rising outside San Diego *(left, bottom)*. Here the University of California is establishing a huge science-oriented campus incorporating the Scripps Institution of Oceanography. And science-based industries, lured by campus research facilities as yet unbuilt, are already moving into the vicinity.

BIG PROJECTS IN PROGRESS
Two major additions to California's science establishment are seen abuilding at left. At Stanford University, bulldozers *(top)* are burying a tube two miles long for use in the U.S.'s biggest atom smasher. At bottom, Nobel laureates Maria Goeppert Mayer and Harold Urey inspect blueprints amid construction at the University of California, San Diego. UCSD will graduate its first freshman class in 1968, soon after the scheduled completion of Stanford's accelerator.

Augmenting the Community

Of all the products of California's science boom, none is more crucial than a human commodity marketed late each spring—new scientists. Commencement ceremonies in the state's 99 degree-giving colleges and universities are now sending forth some 45,000 graduates a year, of whom more than 20 per cent majored in the sciences. California has a vested interest in these graduates: each year it spends millions on the institutions that educate them, and its economy depends more than any other state's on industries with an insatiable appetite for highly trained manpower.

California can take comfort in a survey of Caltech's graduating scientists in 1963. Of the 350 degrees then awarded, 63 per cent were masters or doctorates, each representing at least two years of advanced study. Of the 140 graduates who concluded their education, only a third had gone to high school in the state, yet more than half took jobs in California.

Comparable statistics from other institutions confirm this conclusion: California attracts, trains and employs a disproportionate share of the nation's science talent. While the state needs still more scientists, its current supply keeps the boom going.

A RICH YIELD OF RESIDENTS
Caltech's class of 1964 is joined by orange-caped Ph.D.s in graduation exercises outside Beckman Auditorium. To recruit California's science graduates, high-technology industries in the state offer salaries well above the national average for specialists in many fields.

7
The Bounty of Technology

A PRACTICAL TRIUMPH OF RELATIVITY
Shown during her 1962 maiden voyage, the *Savannah*, the world's first nuclear-powered merchant ship, attests to the link between scientific concept and technological achievement. The theory of relativity made possible the atomic furnace which drives the *Savannah* by converting matter into energy. She can travel 336,000 miles on one 8.5-ton load of uranium oxide fuel.

WHENEVER CRITICS CONSIDER THE IMPACT that science has had upon modern life, they are apt to take one of two strongly opposing views. One credits the scientist for all our comforts and conveniences, and counts on him to add steadily to human health, wealth and happiness. The other blames him for all that seems cheap and tawdry in our industrial civilization, and berates him for helping foster a materialism that has destroyed traditional values.

On one point, however, the rival views coalesce: for good or ill, the impact of science upon our society has been crucial and comprehensive. Through such means as A-bombs and automation, the scientist wields a technological influence that is immediate and direct. Through the effects of his innovations on our social conventions, esthetic tastes and basic beliefs, he wields a cultural influence that is slower and subtler but just as real. These two influences—the one obvious, the other obscure—will be discussed respectively in this chapter and the next.

Technology, in its broadest sense, is the application of knowledge to practical purposes. To be a technologist does not necessarily require either scholarly training in science or the formal title of scientist. Ages ago this was proved by the men who invented the plow and the pulley, the wheel and the windmill, and it is still occasionally reaffirmed by the basement mechanic who contrives an easier can opener or builds a better mousetrap. Modern technology as a whole, however, would have been impossible but for the scientist, and the more sophisticated it has grown the more central has been his role. All of today's important advances are of his making: to cite just a few examples, the computer, the atomic reactor, the laser, the new family of substances called silicones. Indeed, he is so intimately involved with technology that it has come to be synonymous with "applied science."

The scientist's preeminence in the realm of technology stems from his unique way of seeking knowledge—the scientific method—and from the fundamental interplay it creates between ideas and facts. Theories formulated through the method suggest possibilities not only for interpreting nature but also for harnessing it. Newton's conclusions about force and motion made it possible to envisage the behavior of machines as well as of planets. The theory of relativity suggested not only the nuclear processes which keep the sun shining but also the atomic reactors which now drive ships. The concept that atoms are grouped into molecules enabled chemists to understand all the countless substances on earth as varying combinations of the 88 natural elements; at the same time it opened up the vista of new substances that could be made by man himself. In short, ideas generated through the scientific method are confirmed as much by the applied scientist who shows that they work in inventions as by the basic scientist who shows that they work in nature.

That applied and basic science represent two faces of the same coin has long been recognized. As far back as the 14th and 15th Centuries, when the word "science" first cropped up in English writings, it was taken to mean either theoretical scholastic knowledge or skill at a craft. When the natural philosophers of the 17th Century began to practice science as we know it today, they assumed, without question, that technology was to be one of their prime concerns. "The business and design of the Royal Society," wrote its first curator, Robert Hooke, in 1663, "is to improve the knowledge of natural things, and all useful Arts, Manufactures, Mechanick practices, Engynes, and Inventions by Experiments,—(not meddling with Divinity, Metaphysics, Moralls, Politicks, Grammar, Rhetorick or Logick)."

In the triumphs of technology since that time, applied scientists have relied increasingly on basic knowledge provided by their more theoretical, more academic colleagues. The earliest steam engine to be put to wide practical use is credited to a Devonshire ironmonger, Thomas Newcomen. But the principle that steam could be used to move a piston in a cylinder had already been demonstrated a generation earlier by the physicist Denis Papin, an assistant to Christian Huygens and Robert Boyle. When James Watt, an instrument-maker at the University of Glasgow, invented his more efficient steam engine, he was guided by the research in latent and specific heat done by a professor at the university, Joseph Black, one of the first great chemical theorists.

Across the Atlantic, ideas about electromagnetism conceived by other basic scientists in England and on the Continent were applied by Samuel F. B. Morse, Alexander Graham Bell and Thomas Edison to produce the telegraph, telephone, electric light and motion picture. Edison's discovery that the filament of an electric lamp gives off particles—later identified as electrons—led to the invention of the radio tube by the Englishman, J. A. Fleming; it also became a cardinal principle of applied electronics. The theorizings of the 19th Century Yale physicist, Josiah Willard Gibbs, about thermodynamics made possible a host of technological advances in metallurgy, high explosives, aviation and refrigeration; every U.S. kitchen with a freezer full of food owes him a debt.

A belated awakening

Curiously, America was slow to wake to the practical value of basic research. Always a land in a hurry to get things done, it had traditionally put stress upon applied science, feeding off theories worked out in Europe and seldom pausing to develop its own. Edison, for one, so scorned ideas for which he personally could see no application that while he patented his discovery about particles—his single contribution to theoretical knowledge—he never bothered to pursue it. Gibbs, now widely

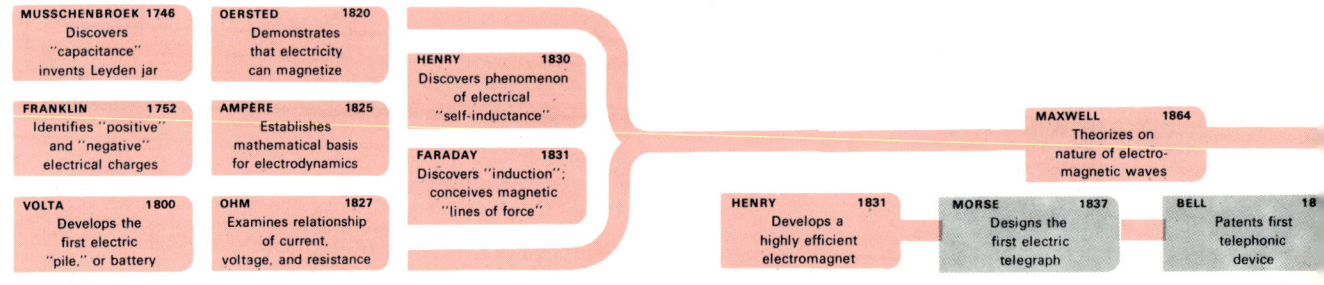

acknowledged to be the greatest native scientist in our history, was a prophet relatively without honor in his own country at a time when Europeans were celebrating his genius. It was only within the past four decades that the U.S. began to recognize that fundamental ideas must be formulated before they can be instrumented, and launched its own intensive programs of basic research.

Today, because of the scope and power of modern theories, basic and applied scientists are closer than ever before. Ideas about matter, energy, forces, waves and particles give direction to investigators or inventors in virtually every field from airplane design to zirconium extraction. For example, when the Army and Navy needed a device for nighttime reconnaissance during World War II, physicists knew the properties of the various rays available in the electromagnetic spectrum, and selected infrared, or "black light," as a suitable invisible search-beam to shine from the G.I.'s Snooperscope. More recently, physical chemists drew on their knowledge of the behavior of crystals to tailor-make an improved nose cone for missiles from a glass-ceramic, Pyroceram, by manipulating its characteristics to make it stronger and more heat-resistant.

Mathematics to the rescue

If no theory exists to cover a need, a new one may be evolved. Among the World War II requirements of the military, for example, was an automatic device that could control antiaircraft fire more precisely. The apparatus desired was one that would direct shells not at an enemy plane itself but at some point in the air where the shell and plane would come together eventually. Mathematicians were called in to find a method of predicting and computing the evasive actions which attacking pilots might take. Out of such problems and their solution grew a whole new branch of mathematics—"information theory," which deals with the communication, processing and utilization of information. As developed after the war by masters of mathematics like Norbert Wiener, Claude Shannon and John Von Neumann, information theory has acquired an increasingly broad sweep. It guides the design of radar defense networks, telephone equipment and computers. Geneticists use it to help explain how the information stored in our genes is utilized to direct protein-building in our cells. Neurophysiologists use it to figure out patterns in the flow of nerve impulses. Social scientists find it serviceable in analyzing a nation's economy.

As more and more scientists are drawn into the world of practical affairs, the lag between the development of concepts and their concrete application steadily diminishes. The idea of steam power hung in the air for 1,600 years, from the time of Hero and the miniature turbine he built to the time of James Watt, whose improved steam engine quickened

THE GENEALOGY OF THE RADIO, like that of most major inventions, includes the work of a line of both theoretical and practical scientists. Early theories gave rise to simpler inventions, until most of the necessary parts had evolved. In the chart which begins at left on the opposite page, the pink squares represent theoretical contributions to radio's evolution, the gray squares subsidiary inventions.

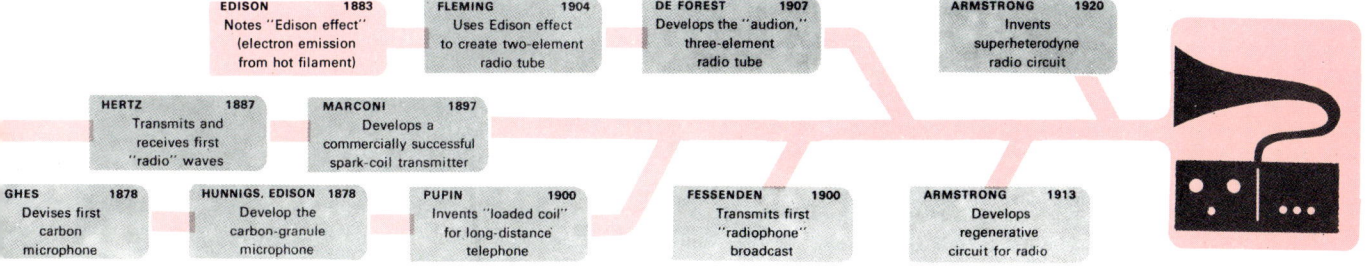

the Industrial Revolution. Photography gestated from 1727, when Johann Heinrich Schulze discovered that silver nitrate darkens in sunlight, to 1839, when Louis Jacques Mandé Daguerre patented the first practical photographic process, daguerreotypy. The telephone loomed as a possibility from 1820, when Hans Christian Oersted discovered electromagnetism, until 1876, when Alexander Graham Bell patented his electric "speaking tube." The vacuum tube was under development for 33 years, from 1882 to 1915; radio for 15 years, from 1887 to 1902; television for 11 years, from 1923 to 1934; the atomic bomb for six years, from 1939 to 1945; the transistor for three years, from 1947 to 1950; and the solar battery for two years, from 1953 to 1955. Scientists joke that if this rate of acceleration could continue, a basic discovery in the year 2500 would turn into a working device in the twinkling of a microsecond.

Idea, invention and impact

The narrowing of the gap between idea and invention has caused a profound revolution in every corner of human life. The layman need look no further for the impact of science on technology than in and around his own home. Most of our conveniences spring from 17th Century discoveries about mechanics, 18th and 19th Century discoveries about heat, 19th Century discoveries about electricity, and 19th and 20th Century discoveries about atoms and molecules. Our carpet sweepers and typewriters may work on purely mechanical principles, or they may be driven by electric motors. Electricity powers our clocks, washers, dryers, sanders, saws, drills, garage-door openers, coffee makers, blenders, ice crushers, toothbrushes, shoe polishers and even back scratchers. Understandings about heat have brought us refrigerators, radiators and internal-combustion engines. Understandings about atoms and molecules have brought us food preservatives, drip-dry clothes, long-playing phonograph records and fiber glass boats.

Abundant as this bounty may seem, it constitutes but a trickle of the flood to come. Such is the forecast of the technological future made not by fanciful amateurs but by scientists themselves. Their reputation for sound prophecy is a persuasive one. As long ago as 1920 they were predicting television sets and computers. Much of the technology that still astounds us, even now, was old hat to them yesterday; much that will astound us tomorrow is matter-of-fact to them today.

Enough eminent scientists have gone on record with lists of the technological achievements they foresee in the next 40 or 50 years to provide the following consensus.

In the home, electrically powered robots capable of moving up and down stairs, and fed with information by the household computer, may run other domestic machines. Illumination will be controlled during the day by windows of light-sensitive glass that automatically darkens toward noon and grows more transparent toward sunset; at night, walls will provide an even glow of electrical lighting until "switched off."

On major throughways, drivers may turn their cars over to automatic pilots directed by master radar systems that monitor the whole road. Commuters will rely on helicopter and hovercraft taxis, and on hydroplane launches skimming high out of the water in port cities. The dangers of air-crowding will prevent the use of private helicopters, and self-piloting will be discouraged except as a sport. In less populous areas, amateur aviators, pursuing the perennial nostalgia for the strenuous life, will test their stamina in pedal planes with enormous wings, propelled entirely by muscle.

Jets traveling at speeds between two or three times that of sound will shuttle back and forth between the antipodes, taking no more than seven hours for each one-way journey. They will rise like rockets from a vertical takeoff position, but will never go as fast as rockets, simply because the earth is too small to make such speed desirable.

The airwaves serving communications will be even busier than the air lanes of transportation. Most of the frequencies now available in open space will have been preempted by television, radio and radar, but new frequencies will be exploited by satellites and lasers. A network of satellites in the sky will make run-of-the-mill home entertainment out of round-the-clock, worldwide television. The video enthusiast will be able to insert a tape into his set and watch as well as hear an old Toscanini concert or a Tennessee Williams play. The more affluent may be able to subscribe to a computer-library and, by telephoning, receive a playback of any audio-visual tape in the archives.

The spider webs of megalopolis

Megalopolis will have a less massive look. At present the materials with which architects and engineers fashion buildings are about a hundred times weaker, structurally, than the potential strength of their molecular bonds makes theoretically possible. As solid-state physicists and metallurgists find commercially practical methods of eliminating flaws and impurities from steel and aluminum, smaller quantities of them will go a longer way, and metropolitan landscapes will be graced with bridges, skyways and spires as airy-looking as spider webs.

The businessman or the diplomat, whether at his desk or at home, will be able to confer with colleagues across the oceans through a videophone hookup. Computers, too, will "confer" with one another through large private networks. Machines that keep track of inventories and payrolls in big business may also swap information with branch-office government computers in return for the lowdown on overall regional,

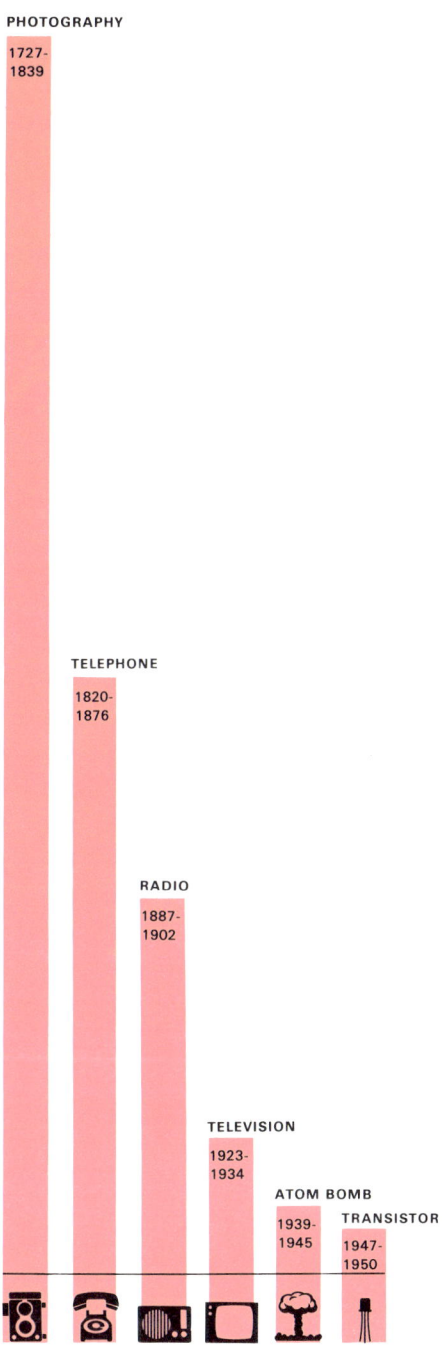

THE TIME SPAN between a major scientific discovery and the first patent to evolve from it has been decreasing. More than 100 years elapsed between the first photographic image and the patent of the process; 56 years between the discovery that electricity created a magnetic field and the telephone. The remaining dates above refer to the first transmission of radio waves and the patent of the radio; the first electronic scanner and the first television patent; the splitting of the atom and the secret patents for atom-bomb mechanisms; and the first observed "transistor effect" in a semiconductor and the transistor.

AIRPLANE SPEEDERS are accosted by a high-flying policeman in this 1911 American cartoon. Although the Wright brothers had made a successful flight eight years before, skeptics still saw the airplane as an improbable toy. In 1947, only 36 years later, the first airplane broke the sound barrier.

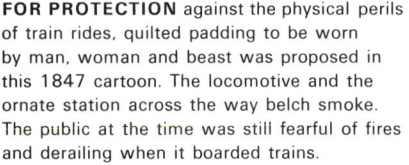

FOR PROTECTION against the physical perils of train rides, quilted padding to be worn by man, woman and beast was proposed in this 1847 cartoon. The locomotive and the ornate station across the way belch smoke. The public at the time was still fearful of fires and derailing when it boarded trains.

national and international economic conditions. The local government computers, in turn, may feed into huge, centralized federal machines, which will be programmed to monitor and predict fluctuations in the national economy, thereby serving Washington as a major tool in the shaping of economic policies.

Elected officials may listen to the voice of the people through computerized polls. By pre-arrangement, interested citizens may telephone in their opinions of a particular piece of proposed legislation, and auxiliary equipment may relay the messages to a computer in the office of their Senator or Congressman. The results of such polling would be known within an hour.

A new look in seashores

At the seashore men and women may watch trains of plastic "sausage skins," filled with cargoes of oil, ore and produce, sinking beneath the surface in the tow of fast nuclear submarines. They may see underwater vehicles descending into the surf on their way to sea-bottom mines, gardens and sport-fishing hostels.

In the laboratories of the geneticists desirable mutations may be induced in plants and animals, and farmers may find it routine to try out new strains of asparaguses and Anguses. In the laboratories of the physiologists and biochemists, there will be talk of lengthening life indefinitely and of improving the I.Q.s of the unborn.

All these possibilities and many more are considered by today's scientists to be both feasible and realizable within a matter of decades. Indeed, some of the projected developments, like light-sensitive glass, pedal planes and submarine dwellings, are already being tried out experimentally.

Science can undoubtedly deliver the promised goods, but there is some question as to how many of them society can absorb in any one span of time. Every new invention spun off by the whirling genie of science makes waves which travel out to every quarter of our civilization. Manners, morals, social attitudes, habits of work and play, ways of war and peace—nothing remains untouched. If a society is forced to adapt to too many changes too rapidly, its fundamental structure may suffer serious strain and stress. Primitive tribes under such circumstances have been known to disintegrate. Civilized men are more supple, but no one really knows how many technological revolutions they can withstand.

Already modern transportation has created a vast gypsy class, which recognizes no hometowns, which has no sense of community belonging, which does not look to neighbors for approbation and respect. Rapid changes in the domestic scene have made it increasingly difficult for to-

day's children to visualize the conditions of life in which their grandparents grew up. Passages in novels written a mere century ago are often incomprehensible except to social historians. The wisdom of the ages, expressed in story and parable, is in danger of becoming a closed book. And future generations are threatened with the task of having to reconstruct for themselves the codes of ethics and esthetics on which a fully satisfactory life depends.

Each new change which engulfs us stirs controversy, and the scientist stands at its center. Inevitably, those who welcome his achievements are less vocal than those who worry over them. The concern is not just recent. Even so staunch a believer in science as Thomas Jefferson, noting the ravages of the Napoleonic Wars, could confide to John Adams in 1812: "... if science produces no better fruits than tyranny, murder, rapine and destitution of national morality, I would rather wish our country to be ignorant, honest and estimable as our neighboring savages are."

A century and a half later, critics see confirmation of Jefferson's unease in the development of weapons capable of annihilating the entire human race. This item heads but by no means exhausts the bill of particulars which they have drawn up against science. They also list urban sprawl, smog and soot; shoddy goods made of artificial substances; a deadening standardization; and the all-pervasive crowding and pressures of modern living.

The subway or the trireme

In rebuttal of this harsh indictment the enthusiasts of science point out that technology has reduced infant mortality, lengthened the lifespan and elevated standards of living. Further, they ask: does technology prevent anything from being beautiful? Is not the Golden Gate Bridge, or Rockefeller Center, or any piece of fine machinery beautiful? Does an "artificial" nylon stocking become a lady's leg worse than a "natural" silk one? Is it more stifling to ride a rush-hour subway than to row an ancient trireme? Is it more immoral to kill by atomic fission than by the sword?

In short, the defenders of technology say that most of the complaints about it are irrelevant. Technology of itself is neither good nor bad; it is what we make of it.

Even so, the scientist is deeply disturbed about the technological power he wields. Hiroshima is graven on his mind, yet he finds himself impotent to control the effects which his discoveries may ultimately have. Clearly, neither Archimedes, who invented the compound pulley, nor Joseph Henry, the first American to devise an electric motor, nor the Otis brothers, who installed the first electric elevator in an American

A ROBOT SETTING TYPE was pictured in an English trade journal in 1884, the year the Linotype was invented. Printers feared the machine might take away their jobs—and for good reasons. At the time the cartoon appeared, a man set four lines of type a minute. By 1945 a man and a Linotype machine could set 15 lines a minute. Today's automated processes can set 1,000 lines a minute, and scientists are predicting that computers will be able to set 60,000 lines a minute by 1984.

office building, can be blamed for the fact that thousands of men and women must daily battle their way in and out of skyscrapers en route to and from the organization life.

Even if the scientist were endowed with some special sort of foresight, he knows that it would be irrational and futile to suppress a find that he thought might be abused in application. The laws of nature are the same everywhere, open to the understanding of men in any land. History has shown repeatedly that a set of equations or experiments which suggests a possibility to one scientist is almost certain to suggest the same possibility to another scientist somewhere else. In the course of the single decade of the 1820s, for example, the electric motor was independently invented by Michael Faraday in London and by Joseph Henry 3,000 miles away in Albany. The pattern was repeated in the 1930s, when Allied and Axis physicists converged on the trail of radar, and still later, when the synchrotron was simultaneously invented on both sides of the Iron Curtain.

The scientist, in short, feels that he has no choice but to keep going, and to hope that his fellowmen will make the best use of the bounty he produces, however mixed it may be. He knows, better than anyone else, that human affairs are increasingly destined to be linked with applied science. Whether for good or evil, that destiny is manifest.

Old Guesses about the Future

The growth of scientific inquiry in the 17th Century produced such an immediate harvest that Francis Bacon soon predicted a scientific Utopia to come *(opposite)*, in which science would make knowledge and abundance possible for all. By the early 20th Century, however, steam and electricity had already wrought revolutionary technological changes, and the public had discovered an ironic truth: the fruits of science are not all necessarily sweet. Everyone was caught up in the new world run by machines, and the popular press was filled with comment on the power of steam and electricity and on the new fields to be explored under the oceans, in the air and in outer space. The cartoonists and illustrators of the period amused their readers with guesses as to what would happen next. Some of their pictures expressed outrage, some optimism, but most expressed a wry common agreement that science was making change a permanent way of life.

A 17TH CENTURY UTOPIA
The scientific wonders *(opposite)*, drawn for this book in mock-17th Century style, were described by Sir Francis Bacon in his book *New Atlantis*, published in 1627. The telescope (d) and microscope (f) were already in use then, but the other arts and tools of his Utopia, like the telephone (n), were all sheer prophetic guesswork. He even anticipated the laser beam (e), invented in 1960.

NEW ATLANTIS

a   Wildfires burning in water
b   Engine houses to study motion
c   Ability to fly in air
d   Instruments for seeing distant objects in the heavens
e   Light intensified and thrown great distances
f   Glasses to see small bodies perfectly
g   Perspective houses to study light and color
h   Pools to strain fresh water out of salt
i   Gardens bearing more speedily than their nature
j   Animals bred both greater and smaller than their kind
k   Fruit much larger than its nature
l   Aids to improve hearing
m   Sound houses for studying sound
n   Sound conveyed in tubes over distances
o   Deep caves for refrigeration
p   Ships sailing under water

L. Hess

"Unconquered Steam" Conquered

*Soon shall thy arm, unconquered steam, afar
Drag the slow barge, or drive the rapid car;
Or on wide-waving wings expanded bear
The flying chariot through fields of air.*

Erasmus Darwin, Charles Darwin's grandfather, wrote these lines in 1791, just a few years after James Watt perfected the steam engine. He proved to be a remarkably accurate prophet. Although steam failed to produce a practical flying chariot, from the 1800s on the power of steam was rapidly put to use on land and water. The changes wrought by the steam engine made it the mightiest economic and social force of the 19th Century. Steam made practical for the first time giant industries and enterprises. Bigger and faster became synonymous with better.

Such a change could not take place without great public outcry: from the landowners through whose property the railroads were run, the farmers who claimed the awful engines would make cows stop giving milk, the factory workers who had to labor longer to keep up with tireless machines that could run 24 hours a day. But fight as they would, the opponents of steam found their adversary unconquerable; by mid-century the civilized world was running on steam.

VÉLOCIPÉDRAISIAVAPORIANNA
This improbable name, which freely translated from the French means "bicycle driven by vapor," was invented by the author of this 1818 French print to describe the equally improbable machine he had conceived. Although someone somewhere may once have mounted a boiler on a bicycle, the artist was not picturing a real vehicle, but simply caricaturing steam power.

THE RUIN OF A CITY
Steam engines out of control, passengers felled by noxious fumes, a city covered with soot—all these were forecast for London by artist Henry Alken in an 1828 engraving called *Illustration of Modern Prophecy*. Steam locomotives had just begun making regular runs in England.

THE ELECTRIC HORSE
In his series of stories about teen-age inventor Frank Reade Jr., Lu Senarens came up with a host of inventions. This cover illustration is from *The Electric Horse,* published in 1896. Electricity had begun turning the wheels of railroad locomotives the year before, but Senarens did not make the jump to the horseless carriage, inventing instead an unwieldy electric horse.

The New Life with Electricity

By the 1870s people were becoming accustomed to the steam-powered world. Now, however, right on the heels of steam came electricity. Soon yokels at country fairs were eagerly paying for the thrill of a harmless shock from a storage battery, while more sophisticated city folk were strolling down arc-lighted avenues and buying electric hairbrushes, electric corsets and magnetic belts. The telephone came in that decade too, and city skies were soon filled with wires.

Popular novelists anticipated such never-never contraptions as electric horses *(left)* and machines to end books *(below)*. They also foresaw with startling accuracy such innovations as TV *(bottom, left)*, and many other inventions we take for granted.

THE END OF BOOKS
In 1895 the French artist Albert Robida prophesied that bibliophiles would one day be replaced by phonographophiles *(above)*. No more space-consuming books—merely turn on the "Universal Phonographic Library," set the dial to "novel" or "philosophy," lie down and listen.

THE TELEPHONOSCOPE
In his 1883 book *The Twentieth Century*, Albert Robida described the Telephonoscope *(left)*. To the telephone, already in use, he added a viewing screen. A turn of the dial and subscribers all over the world could watch any current theatrical production or tune in on a friend at home.

20TH CENTURY PARIS
Electric aircars, billboards advertising a thousand products and buildings of glass were all part of the Paris Albert Robida predicted. In this illustration from *The Twentieth Century* the concierge is accepting the card of an airborne caller. The skies are filled with fliers—and wires, which Robida predicted would often electrocute or even decapitate the unwary passenger.

ELOPEMENT, AIR-AGE STYLE
As papa pursues in his earthbound auto, waving his cane in rage and frustration, daughter swoops romantically off in the airplane of her fiancé *(right)*. The airplane in this 1911 drawing may look farfetched, but when it appeared in the humor magazine *Life,* flying machines still looked somewhat like bicycles with wings.

Bicycles, Flivvers and Tops Aloft

On December 17, 1903, after centuries of dreaming about flight, the world awoke to the air age. On that day the Wright brothers first flew their plane. But press and public, weary of earlier false claims of successful flight, took little notice, and remained vastly uninformed even about the shape of the new aircraft. Cartoons of the period envisaged it built of box kites and bat's wings *(opposite, below),* and prophesied crash landings of aerodynamically impossible monsters *(left, below).* So little fame attended the birth of the airplane that in 1906 a Brazilian, Alberto Santos-Dumont, took off into the air in France and, with no knowledge of the Wrights' achievement, thought himself the first to pilot a heavier-than-air craft. The air age had begun twice, on two continents, and still was ignored.

The air aces of World War I convinced the public that the airplane was to be taken seriously, but even in 1920, the gimcrack look of it was a joke. Cartoonists foresaw future craft shaped something like a top—with a headlamp in front and a lunch basket on one wing *(opposite, top).*

MAROONED ON A MOUNTAIN
The title of this 1909 illustration is "The 'Transcontinental Flyer,' Lost in Storm, Stranded on a High Rocky Mountain Peak." Like the cartoon above, this picture is by the American, Harry Grant Dart. Unlike most journalists of the time Dart was fascinated by the airplane, filled *Life* with imaginative ideas of airborne craft to come.

"Ta ta, Father"

A LATTER-DAY WHIRLWIND RIDER
Another Dart creation, "The Elijah" *(above)*, the imaginary private airplane of the future, took its name from the Old Testament prophet who rode a whirlwind. Dart's cartoon, patterned after contemporary automobile advertisements promoting the new models, appeared in 1920.

THE LAST AUTOMOBILE
This 1907 cartoon, which appeared a year before Henry Ford launched the Model T, forecast the death of the automobile by 1950 *(below)*, as families took to the skies in the Tin Lizzie's air equivalent. Gas stations would be hovering in the air to refill the tank of an improvident flier.

IN 1950.
"Why, there's an automobile! How funny it looks!"
"Yes. That's old fossil Jones—says he can't stand these newfangled notions."

TRANSOCEANIC RAPID TRANSIT
In this 1913 cartoon *(above)*, an embarkation for Europe by underwater cable car includes late arrivals, parents hauling children, a fanged fish and underwater fishermen using worms.

Invading the World of Neptune

When the prognosticators of the early 20th Century turned their minds to the ocean's depths, they imagined few of the riches we now know to be there. Instead they merely elaborated on devices already invented for staying under water. The submarine had been around as a visionary craft for nearly 150 years—since an American, David Bushnell, launched a turtle-shaped underwater vessel in 1775 —and diving gear was well developed. The cartoonists might have taken the undersea world more seriously. There was precedent enough in the fiction of Jules Verne. In 1869, in *20,000 Leagues under the Sea*, Verne not only accurately described an electric-powered submarine, he also showed an awareness of the ocean's riches. The *Nautilus'* Captain Nemo smoked cigars made of seaweed, stocked his larder with fish and sea plant life and was able to coax oysters to produce giant pearls. In fact, exploration of the ocean proved more difficult than its proponents had thought, and it was not until after World War II that the public began to hear of serious scientific plans for mining the sea.

THE UNIVERSITY SUBMARINE BOAT-RACE, A.D. 1950.

FULL AHEAD FOR ALMA MATER
Bubbling with enthusiasm, British spectators in a 1913 *Punch* cartoon cheer their man on. Here, as in many underwater cartoons of the period, the joke consisted solely of transferring a landlubberly way of life to the bottom of the ocean—the artist even went so far as to provide dress-shaped diving suits for women, straw hats for the racers and a helmet for the bobby.

THE HECTIC UNDERWATER LIFE
A busy world under New York Bay is shown in a cartoon of 1910. Guests at the Neptune Hotel dance or play tennis; sportsmen in diving suits shoot fish; sight-seeing submarines tour the area; there is even a public library *(lower center)*. The unlucky birds, suspended from a family bungalow, are about to drown, their air tube snapped by a recklessly driven sub.

And Then, Out into Space

The public of the 1910s and 1920s heard more about space travel than about any other promise of technology. Thanks largely to Hugo Gernsback, who founded a string of science-fiction magazines, space fiction was a booming literary genre decades before man reached space. Most early space stories were rife with BEMs (bug-eyed monsters) and eye-filling heroines—but Gernsback's respect for scientific fact and his knack for accurate prediction often gave his pages a solid scientific content. The illustrations of his colleague, Frank Paul, set the standard for the field: fictional space hardware had to grip the reader with its inventive detail and credibility *(opposite)*.

Fact has by now caught up with much of the gadgetry of early science fiction. To keep one-up, writers in this field have had to turn to more thoughtful consideration of the impact of science on the future of man. Many see destruction looming. But for some, the scientific Utopia Bacon dreamed of glimmers in the distance.

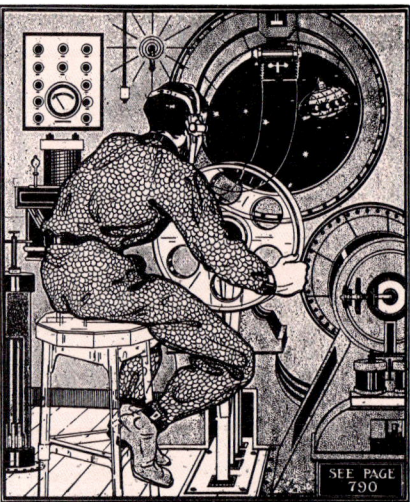

PURSUIT OF A BAD GUY FROM MARS
Ralph, the hero of a 1911 novel by Gernsback, hurtles through outer space in his Space Flyer *(above)*, on the trail of a Martian who has kidnaped his sweetheart. The novel, which was serialized in Gernsback's first magazine *Modern Electrics*, predicted radar in accurate detail.

ATOM-SLICING RAY
Sliced in two by a powerful ray which disintegrates atoms *(left)*, a severed spaceship hurtles toward the ground as men and supplies tumble through space. The attacker's ray has also sliced off a piece of the tail in this 1929 illustration by Frank Paul from *Air Wonder Stories*.

SATELLITE IN SPACE
A space-traveling couple, wearing goggles to protect their eyes from sunlight unfiltered by atmosphere, look out on a 1929 conception of a satellite. This picture by Paul illustrated a three-part satellite community consisting of living quarters *(center)*, an observatory *(left)* and solar powerhouse. Weightless electric cables extend from the powerhouse to the other two.

8
The Impact of Science

BY BOTH WORD AND DEED, the scientist has come to influence our public affairs, our esthetic standards, our religious beliefs, our habits of thought—indeed all the elements of "culture" in its fullest sense. Every major controversy and conversation of our time concerns him: from disarmament to delinquency, from urban blight to birth control, from the space race to the smoking peril. No less than the humanist, he is preoccupied with problems of democracy and morality, with the nature of beauty and truth. He is, in short, a live and electric presence in the world of ideas, showering sparks in all directions.

The root cause of the scientist's deepened impact is the revolution he has wrought in our fundamental views of reality. Astronomers and physicists have discovered a universe of galaxies and atoms that challenges earlier notions of the world around us, and far outstrips the imaginings of past poets and philosophers. It is a universe in which man finds himself discouragingly small and uncentral. At the same time it is one which offers him virtually unlimited scope in space, and a future in time estimated at about five billion years. The scientist has faced up to the immensity and the challenge of this prospect with uncompromising realism. Inevitably his attitude has colored the thinking of other men of the 20th Century. It has helped make them less orthodox and more skeptical than their forebears, more receptive toward new ideas.

All up and down the line, by means of the specific knowledge he has developed, the scientist has moved in on realms that humanists long thought to be exclusively their own. History, once accepted on the written word of our ancestors, is being revised by the radioactive dating techniques of the isotope chemist. The mysteries of the brain are beginning to be plumbed by the biophysicist. The hunch of the politician has been supplemented by the probabilities of the pollster. Computers are programmed to distinguish between literary styles, to compose music and to perform many other feats once deemed impossible without the exercise of human creativity.

Not surprisingly, many humanists dislike these inroads by science. They fear that life is being depersonalized, art vulgarized and religion deposed. To be sure, few of them seriously maintain that the old days were better—at least not for the galley slaves, the peasants or the hosts of children who never lived to grow up. But the churchman misses the sturdy faith of the peasant. The statesman looks back with nostalgia at the personal finesse that largely shaped public policies in the past. Many artists yearn for the small select audiences that set taste and provided patronage for their bygone colleagues. And all humanists regret the loss of great versatile talents, the potential Da Vincis or Goethes of our age, who may now be channeled into the specialized pursuit of science.

So long as the scientist remained, as it were, on his own side of the

THE ARTIST AND DR. EINSTEIN
The fascination that science holds for today's artist is seen in this lithograph, *Relativity*, by Maurits Cornelis Escher, a Dutchman. In this dizzying arrangement each faceless individual moves as if he, at least, knows which end is up, and each is right, according to Einstein's theory, which holds that position in space is always relative to the reference frame of the beholder.

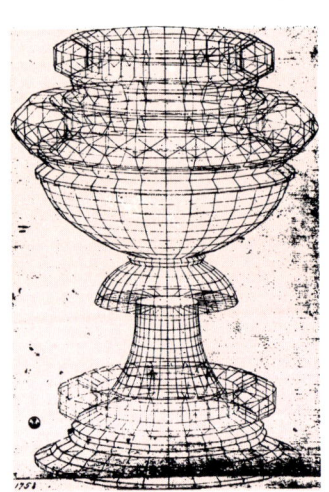

A LESSON FROM SCIENCE is shown in this drawing of a chalice by the 15th Century Florentine painter Paolo Uccello. The perspective of medieval paintings was frequently primitive, but during the Renaissance, Italian artists eagerly experimented with newly discovered rules of optics, showing with mathematical exactitude objects seen in space from a single point of view.

fence, those on the other side delighted to borrow and dabble in his ideas. Dante and Milton both drew upon contemporary lore about astronomy. Marlowe dramatized a prototype of the scientific character in his rendering of Faust, the legendary folk figure who would hazard all for understanding. Shelley once described outer space as a "black concave," expounding on his poetic phrase in a casual scientific footnote which explained, more or less correctly, that the reason the sky was not also black, but blue, was "owing to the refraction of the [sun's] rays by the atmosphere." Haydn wrote an opera, *The Apothecary*, celebrating the pharmaceutical chemist of his age. A lesser composer of the 18th Century, Marin Marais, produced a sonata for viola da gamba about a gallbladder operation, an opus regrettably muted through the mists of time.

Today the claims of science on our culture have become so insistent and pervasive that practitioners of the arts can no longer pick and choose among them. They feel compelled to take a stand either for science or against it. Shostakovich musically re-creates the clangor of a factory in order to glorify scientific progress. Yeats "purifies" his poetry of normal syntax in order to defy scientific logic. Opposed or not, there is little doubt that science, or the machines of its handmaiden, technology, have evoked the intellectuality and austerity of much of today's music and poetry, art and architecture.

In modern art alone, the response to scientific discovery has been irresistible. An eminent art historian, the Frenchman Marcel Brion, has observed, "... new aspects of nature, rarer, more surprising, more varied, testifying to the majestic beauty of cosmic life, arouse the painter's emotions even though he may have no scientific training. ... By ... vicarious use of the microscope or telescope, the artist becomes intimately acquainted with a nature different from the one known by his predecessors." The findings of science have spurred whole styles, including Futurism, which exalted the dynamic of motion, and Cubism, which saw the reality of nature in geometric terms.

The emancipation of color

The impact of science on painting has been both technical and philosophical. Until about a century ago, for example, the artist regarded color solely as a qualitative value, immune to scientific analysis. Eventually the scientist learned to discriminate between the hue, brilliance and saturation of pigments. He thus converted the elusive qualities of color into measurable quantities, and so provided greater precision and range of choice for the painter's palette. Color for its own sake has since come to be a major consideration in painting; often, for example, the flaming canvases of a Kandinsky and other Abstract Expressionists are primarily exercises in color.

The philosophical implications of scientific discovery have had an even more far-reaching effect. The scientist has constantly found that things are not what they seem; that behind every curtain is a new and closer approximation of reality more abstract and unexpected than the last. Buoyed by this approach, the artist, too, has tried to penetrate beyond the surface façade. Familiar shapes dissolve. A Cézanne mountain becomes a heap of planes and angles. A Picasso woman becomes a rectilinear schizoid. Miro's people become curious "biomorphic" creatures which are neither fish, fowl nor human. Mondrian's early conventional landscapes give way to stark arrangements of precisely placed horizontal and vertical lines. "For the modern mentality," notes this master of abstractionism, "a work which has the appearance of a machine or a technical product increases its artistic efficacy."

Echoes of a rallying cry

In architecture, science has left an indelible imprint by way of functionalism, the idea that a design cannot be beautiful unless it serves its intended purpose. This credo grew out of the astonished realization by 18th Century scientists that the beauties of nature all appear to reflect some underlying usefulness and order. The pre-Darwinian French naturalist Jean-Baptiste Lamarck, in propounding an early theory of evolution, developed the now-famous concept that "form follows function." Although Lamarck was applying it to living things, it was destined to become a rallying cry in the world of art. In 1852 the American sculptor Horatio Greenough put forth the thesis that when function changes, artistic form must change as well, and that, further, old forms cannot be used to embody new functions. Ultimately this philosophy led to the clean-lined towers of the skyscraper pioneer, Louis Sullivan, to the earth-hugging houses of Frank Lloyd Wright, and to the scrupulously ornament-free steel and glass structures of the Bauhaus school. More recent and more decorative edifices—those billowy museums and pavilions which have begun to add a new baroque touch to U.S. cities—also owe their existence to science, through both stronger materials and improved methods of calculating the amount of strain and stress concrete can take when shaped into "free forms."

The world of music, too, has increasingly heeded the voice of science. Whatever a Mozart might have made of it all, today's audiences scarcely blink at compositions entitled *Ectoplasme, Tropism I, Location in Space,* and *Density 21.5.* Opera-lovers weaned on the arias of *Tosca* and *Lohengrin* may now listen to librettos on the lives of Kepler and Einstein. The modern musician is enthralled not only by the subject matter of science but also by its tools and techniques, especially those born of mathematics and electronics. The Greek composer Iannis Xenakis programs his themes into a computer, plays and mulls the print-outs, and upon this base builds works which he describes as "above all the expression of a modern symbolic logic." The German Karlheinz Stockhausen composes "electronic music" by blending, to the desired mix, the sounds produced by electronic devices called signal generators.

On occasion, however, experimentation with the sounds of science turns up proof that they are by no means automatically conducive to creative interpretation. For example, the late Paul Hindemith, who ranks among the musical giants of our century, once became intrigued with the spectrum of the hydrogen atom. He asked one of the authors of this book to translate this spectrum into acoustic frequencies. Hydrogen, the most plentiful and probably the most primeval of the known elements, radiates energy at a number of distinctive frequencies, and their use in a melodic theme seemed to promise a truly cosmic composition. When this concept was tested, however, the result was a "tone row" in which a great many high notes, hardly differing from each other, were all crowded together at one end of the acoustic spectrum, while the rest of the octaves were left almost blank. Hindemith forthwith abandoned the idea as musically sterile.

In trying to assess the eventual impact of science on the arts, most scientists and artists alike reject the notion that science can ever dominate the arts completely. The one is a domain of intellectual realization, helping man to know clearly; the other is a domain of the evocative and the emotional, helping man to feel deeply. While both draw upon creative inspiration, each has its own proud jurisdiction, and any serious attempts at trespass would be futile.

The future of creativity

There are, however, particular and important ways in which science is likely to continue to exert an influence on the artistic world. One strong probability is an even greater reliance on scientific tools and techniques as aids in the creative process. Future painters may do their preliminary sketches in color on electronic "slates." Choreographers may animate stick figures by push button. Authors may employ mathematical analyses which warn that a work in progress suffers from too much repetition. Playwrights may utilize geometrical theories of plot structure. The theatrical producer with a new show for Broadway may find out-of-town tryouts more costly and less instructive than the use of a closed-circuit network that tells him precisely when his audiences laugh, gasp or tune out. However fantastic, if not repugnant, many living artists may find such ideas, their successors will almost certainly prove a more receptive breed. They may create in strange, untraditional media, but there is no reason why the best of them should not inspire and

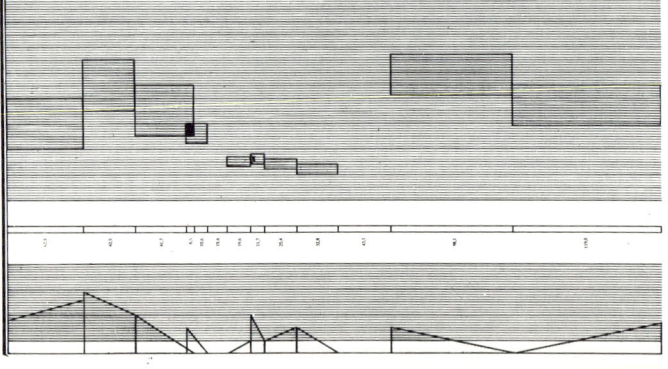

MUSICAL SCORE by the contemporary German composer Karlheinz Stockhausen shows the scheme of notation used for recording his electronic music on tape. The 81 lines in the upper half of the score provide a frequency scale for pitch. Bars indicate the pitches to be produced by electronic equipment. The segmented center line indicates the duration of each sound in terms of centimeters of tape. The lower portion of the score is a decibel scale, showing the volume at any given moment.

entertain as fully as did a Bach or a Shakespeare in times gone by.

Along with the tools and techniques of science, the artist will increasingly embrace experimentation. Intimations of things to come appear even now in the painter who approaches his canvas not with brush but with air gun, and the sculptor who foregoes marble for junk iron. Indeed, there are those who argue that experiment for its own sake becomes an act of art. This avant-garde of the avant-garde includes the sculptor who bends wire while engaged in uninhibited ballet, and the composer who leaves everything to the last minute and confronts his audience with an impromptu "event" in lieu of a recital.

A box-office benediction

Today's public is far more receptive to serious innovation than in the days when the dissonances of Stravinsky's *Rite of Spring* provoked riots in concert halls. It amiably tolerates even the outright exhibitionist shenanigans. Indeed, by bestowing the benediction of the box office, by gallery-going and art-collecting and record-buying to an extent unprecedented in years past, the public has provided an earnest of its desire at least to understand the new, if not necessarily to endorse it. Science may have produced one of its finest harvests in a mounting popular willingness to keep an open mind, and in a corollary refusal to cling exclusively to yesterday's esthetic criteria.

It is unlikely that even the best-informed layman is fully aware of the part science has played in helping mold his tastes, and equally unlikely that he realizes the scope of its role in shaping his civic and economic affairs. The influence of the scientist on government is considerably more direct than on the arts, yet most citizens, if questioned on this score, would describe his participation as largely limited to the sphere of nuclear policy and weapons research. In fact, however, he has made his voice heard and his presence felt in virtually every federal agency and office.

With a White House declaration of a war upon poverty, social scientists are marshaled to roll out the statistics and spot the points of attack where money can do the most good. Testimony by biologists helps conservationists beat back powerful ranching and mining interests and push through a Wilderness Bill that sets aside land where "the earth and its community of life are untrammeled by man." Economists monitor and advise on the financial state of the nation through mathematical models of the economy by which they analyze the implications of stock prices and other crucial indicators.

The extent to which we are being governed scientifically can be best appreciated in terms of the sheer volume of data on which the government now bases its decisions. In the last decade alone, by making it possible to automate clerical work, the scientist has increased by several hundredfold the amount of factual information available to public servants about our country and its people. It has been estimated that if all this information were being amassed manually, the entire U.S. population, working full time, could not even begin to keep up with the paper work involved.

The scientist's influence on public affairs is most discussed, and most decried, in the use of polls and sampling techniques. Critics assert that candidates for high office make their decisions to run or not to run on the basis of survey rather than on the conviction that they can serve effectively; that once in office they are unduly guided by the dictates of statistics and committees of experts, rather than by their own instincts for the needs of the people.

Many scientists dispute these views. They see the polling technique, implemented frequently and almost instantaneously through electronics, as an adjunct to the voting process. They regard both the committee system and the computers that keep track of growing mountains of fact as blessed aids to administrators in decision-making. Some of them, indeed, seriously harbor a dream that the cantankerous citizen will find farfetched if not downright appalling. These dreamers foresee the possibility of a single, unified, democratic world government that receives its guidance not only through the ballot box but also through careful soundings of public opinion and scientific analysis of local conditions. If computers were well utilized, information specialists believe, a world president could be better informed about his constituents than any provincial governor or town mayor of the past.

The sweep of an indictment

Of all the broad influences of science, the most difficult for many of us to accept are those affecting religion. The reason is clear. In matters of government and the arts, we tend on the whole to see the scientist as a constructive force. In matters of our cherished beliefs, we are apt to see him as totally destructive. Some of the most illustrious practitioners of his craft have come under indictment on this count. Copernicus is said to have bereft man and his planet of their central place in the universe. Darwin is said to have demoted man to the ranks of the animals. Freud is said to have robbed man of his soul. Today's neurologists, biophysicists and biochemists are said to have reduced the human brain to a machine.

Scientists have a ready retort to these assertions. As they see it, Copernicus formulated an improved and more wondrous model of the universe; Darwin revealed the remarkable process by which man came to rise above the lower animals; Freud uncovered new depths and complexities in the human mind. None of these men, say the scientists, was passing any kind of judgment on religion; they were simply revealing the scope of creation, and their revelations have had little bearing on man's relationship with his God—except to heighten his sense of awe.

Curiously, in view of the protracted conflict between them, theologians and scientists generally agree—and philosophers concur—on the nature of religion. They hold that it consists of three principal components. The first is faith, and specifically a faith that the universe and mankind were created for a purpose. The second is ethics, the principles of human morals or right conduct. The third is explanatory doctrine, devised by ecclesiastical thinkers to expound on the relevance of ethical behavior to the purposes of creation.

Bridge-builders and a battle

It is these metaphysical doctrinal bridges which have caused the trouble between science and religion. The theologians who constructed them declared, for example, that the stars needed seraphim and cherubim to keep them in motion; that the place of man was at the very center of the universe; that he was created in one day. The scientists came along with their own unsettling denials of such pronouncements and the battle was joined. Martin Luther, for example, denounced his contemporary, Copernicus, with a harshness that was unalloyed: "People give ear to an upstart astronomer who strives to show that the earth revolves, not the heavens, the sun and the moon.... This fool wants to reverse the entire science of astronomy; but sacred scripture tells us that Joshua ordered the sun to stand still, and not the earth."

The discoveries of Copernicus, however, took man only a step away from his blissful belief that the world had been created around him, that his abode in the universe was fixed and central. He could still believe, while conceding the majesty of the universe, that he was unique —God's prime creation. But as science has persevered, even this consolation has dwindled. There is every reason to suppose—we are told by the successors of Copernicus—that billions of stars have planetary systems, and that myriads of them are suited to sustain forms of life. Our last conceit is threatened; it is very likely that man is not the crown of creation, and that there are innumerable beings with intelligence perhaps greater than our own. For our sun is a fairly young star, our earth is still younger, and if intelligence takes eons to develop, other planets have had an advantage over ours in the time they have had available for evolving life.

Immense and staggering philosophical questions thus arise; they are questions not only for churchmen but for all of us. Are distant creatures moral beings, crushed as we are by a sense of inadequacy and sin? Was

Estimated population of stars:
100,000,000,000,000,000,000.

Of these, 1 in 100 is a single star:
1,000,000,000,000,000,000.

Of these, 1 in 100 has a planet system:
10,000,000,000,000,000.

Of these, 1 in 100 has sufficient atmosphere:
100,000,000,000,000.

Of these, 1 in 100 is neither too hot nor too cold:
1,000,000,000,000.

Of these, 1 in 100 has air, water and land like earth's:
10,000,000,000.

Of these, 1 in 100 has life already under way:
100,000,000.

LIFE ON OTHER PLANETS seems more than likely to exist, according to the educated guess of astronomer Harlow Shapley. Double or multiple stars that gravitate around each other do not have planets, but even when these are excluded from consideration, billions of billions of single stars remain. By giving each successive condition for life—from a planetary system to water—only a 100-to-1 chance, as shown here, Shapley hypothesizes that 100 million stars may have planets possessing living organisms. He considers it improbable, however, that there is an exact duplicate of man on any other planet.

there, perhaps, a Moses, or a Buddha, or a Christ on faraway planets? How does the believer face the prospect of universal rather than global salvation?

Echoes of the centuries-long war between church and science still reverberate. For the most part, however, 20th Century leaders of the major denominations, as well as many scientists, have come to avoid the age-old clash and clatter. Julian Huxley, the distinguished English biologist, has written, ". . . it is no longer possible to maintain that science and religion must operate in thought-tight compartments or concern separate sectors of life; they are both relevant to the whole of human existence. The religiously-minded can no longer turn their backs upon the natural world, or seek to escape from its imperfections in a supernatural world; nor can the materialistically-minded deny importance to spiritual experience and religious feeling."

The decline of dogmatism

Both sides have yielded up a goodly measure of their own brands of dogmatism. Both have relinquished needless pretensions about the certainty of their respective metaphysical and physical models of the universe. The theologians have bowed to historical perspective and to the conviction that faith and morality, rather than metaphysics, are the key elements of religious experience. The scientists have bowed to recent and tremendously important findings by science itself.

During the early decades of the 19th Century many scientists believed that the universe ran as a machine; that its parts consisted of colorless, odorless, silent masses of moving matter. Given complete knowledge of the machine at any one instant, the French astronomer Laplace argued, a mathematician should be theoretically able to work out and predict all future states of the machine and all future events. This mechanistic, deterministic attitude stemmed from common-sense observations of the universe which had already brought science far along on its triumphant march.

In the second half of the century, however, mathematicians and physicists began to acknowledge that they could conceive of some aspects of the universe only in mathematical formulae. Out of these came a number of ideas which plainly flouted common sense: space that is curved, matter that may sometimes turn immaterial, energy that possesses mass. Shortly thereafter, Werner Heisenberg and his fellow quantum physicists discovered that in the world of the atom cause and effect are not mechanically linked; that a cause may have more than one consequence, each governed by probabilities; that no one could predict that one course of events would follow another except as a calculable likelihood.

In short, the idea that the universe is a machine and the idea that it operates in deterministic fashion both died within a short time of each other. Heisenberg's "principle of uncertainty" may even have reinforced the fundamental human predilection for free scope: it allows room for chance in the workings of the sex cells where our children are blueprinted by genes, and the nerve cells where our minds are made up and our imaginations fired. The Heisenberg principle says nothing about freedom of choice—the ability to choose among chances—but at least it liberates us from the 19th Century concept that our destinies are completely predetermined.

The downfall of common sense in atomic matters was hard for many scientists to take. Einstein himself was never completely reconciled to the idea that probabilities govern cause and effect. "I cannot believe that God plays dice with the world," he once commented indignantly. Today the physicist believes that in a manner of speaking God does, indeed, play dice. That is to say, the universe is not a machine, at least not in any ordinary sense, and its future is not wholly ordained by the position of its constituents at any one moment. Moreover, after the revelations of recent decades, the scientist's understanding of the universe is regarded by the scientist himself as perpetually open to improvement. Current views of the universe are merely the best and latest models known. Improved models, it is agreed, will increase the scientist's sense of sureness and his ability to predict, but will never provide him with absolutes.

Portents of a "new faith"

Through the discoveries of modern physics, the scientist has found a fresh attitude toward his work. He has acquired a new diffidence about attaining ultimate truth, or putting too much trust in mental images, or reasoning too positively about cause and effect. This growing humility—which may indeed be described as a "new faith"—promises a closer rapport between science and the humanities in the years ahead. Increasingly, thinkers in every field are coming to recognize the merit of the scientific method. Eventually every man, in a sense, may become a scientist, each using the method according to his own interests and needs.

But, ultimately, who knows what men may become? Bernard Shaw envisioned them, under the aegis of science, evolving into vortices of pure intellect. Some science-fiction writers have seen us whirling off as disembodied dervishes to spend our vacations on distant planets. The scientist, in his meteoric rise, has made almost anything seem at least remotely possible.

We have, up to now, barely scratched the surface of understanding.

ROSES

Few fingers go like narrow laughs.
An ear won't keep few fishes,
Who is that rose in that blind house?
And all slim, gracious, blind planes are coming,
They cry badly along a rose,
To leap is stuffy, to crawl was tender.

CORSETS

Yes, illiterate is its rowdy, black is his avenue,
Mine is a hay of these dwarfs.
Does he look like a sin of alabaster?
Moreover, food tastes like coy buttermilk.

STEAKS

Is that the automaton that smells like the tear of grass?
All blows have glue, few toothpicks have wood,
Direct a button but I may battle the ham,
The crafty carnival's kite daintily massacres the scalp.
Yes, we would, you shall,
Shall not I tighten a moose's parasite?

COMPUTER POETRY like the examples above is turned out by a computing machine known to experimenters as A.B. The machine chooses words and grammatical patterns at random. When programmed with 32 grammatical patterns and a vocabulary of 850 words, A.B. produced "Roses." Using a larger vocabulary and more sentence structures, it composed "Corsets" and "Steaks." Although any meaning in these poems is purely accidental, the experimenters hope to program A.B. eventually to write meaningful sentences.

We do not know, scientifically, the purpose of the universe. We do not know, except by instinct and by trial and error, how to get along together, or how to govern ourselves. We do not know what it is about great works of art, or thoughts of any sort, that excites us and makes us press forward. In 6,000 years of recorded history, and perhaps 100,000 years or more of spoken thought, we have only begun to realize ourselves as intellectual beings.

For his part, the scientist has been recognized as a unique force in our midst, and has indeed enjoyed his very name, for little more than a century. As yet he does not understand the workings of a worm as well as he does the movements of stars or electrons. But what will he have achieved for us hundreds or thousands of years hence? The British physicist and Nobel laureate, Sir George Thomson, speculated a decade ago: "Art, religion, patriotism, humanitarianism, what will each look like when we know what circuits are excited, and in what sequence, in the brains that feel the emotions?" Even now, the scientist has immeasurably expanded our horizons. Unfathomable dimensions in both time and space are now open to us. The scientist's influence on culture, technology, communications, scholarship and knowledge has only begun to be felt. And his bright vision of the future—of a world free from ignorance and want—encompasses us all.

The Nobel Prize: Accolade for Greatness

More than most people, scientists pursue their calling for its own sake. The public acknowledgment for a lifetime dedicated to research may be a footnote in a scientific journal. But for those who change the face of science through their work, there may come a moment in the world spotlight. These are the Nobel laureates, winners of prizes in physics, chemistry and medicine. The Nobel prizes are not only prestigious but carry an award that may go as high as $50,000, tax-free. So meticulous is the selection process that scientists have seldom disagreed over the choice of the more than 200 who have been picked since 1901 *(Appendix, page 189)*. The winners are chosen according to the instructions of Alfred Nobel, the Swedish inventor whose fortune made the awards possible: "no consideration be given to the nationality of the candidates . . ." and those chosen "shall have conferred the greatest benefit on mankind . . ."

THE PINNACLE OF A CAREER
Holding the medal and diploma he has just received as the tangible evidence of his scientific achievement and prestige, Otto Paul Diels pensively studies his medal during a pause in the Nobel Prize award ceremony in the Concert Hall at Stockholm, Sweden. In 1950, the 50th year of the awards, he and Kurt Alder *(second from left)* shared the Prize for chemistry. Diels was then 74.

ALFRED NOBEL AT 60
When this photograph was taken in 1893, three years before his death, Nobel was in failing health but still busy. He worked long hours, on both research and business affairs, adding another steelmaking plant to his empire that year.

THE MANSION AT SAN REMO
In 1891 Nobel, shown here with his assistant, Wilhelm Unge, moved to this mansion near San Remo, Italy. He first called his new home *Mio Nido,* "My Nest." But when a friend jokingly reminded him that a nest was for two people, not one, bachelor Nobel renamed it Villa Nobel.

THE LABORATORY AT VILLA NOBEL
On the spacious parklike grounds surrounding his villa in San Remo, Nobel built a gleaming laboratory. During his lifetime he obtained more than 300 patents. Some were in areas closely related to explosives, the foundation of his fortune. But he also tried to anticipate the future by experimenting in other fields, including substitutes for rubber, leather and silk.

Humanitarian and "Death Merchant"

Alfred Nobel, who revolutionized warfare by inventing dynamite, came to have an interest in the field of explosives in the most natural way: his father was in the business. Trained as a chemist, Alfred was a brilliant researcher and a prolific inventor. He perfected dynamite early in his career, and it became the base for an empire which eventually included factories in 11 countries. Nobel was primarily a scientist and a superb technician who disliked the world of finance and trade. He never had an office, but conducted all his business from the laboratory he set up wherever he lived—France, Italy or Sweden. Shy and deeply engrossed in his work, Nobel never married.

Often characterized as a merchant of death, Nobel was a humanist, a patron of the arts, an internationalist and pacifist. Partly because he believed that "no happiness goes with inherited fortune" he left nothing to his relatives. He willed his entire fortune to be invested, the interest to be awarded each year as prizes in science (in the fields of physics, chemistry and medicine), literature and peace—an appropriate bequest by a man whose scientific skill led to deadlier warfare, but was himself fond of the arts and a devotee of peace.

A WALL OF FIRE FOR RUSSIANS
In 1844 Nobel's father Immanuel demonstrated his land mines for Russian military officers. Afterward, he painted this picture of the scene, in which he appears as the short, top-hatted man in the foreground. The Russians were impressed, and bought the weapon. During the Crimean War, an underwater version of the Nobel land mine was used against the British fleet.

The Painstaking Task of Selection

In his precisely explicit will, Alfred Nobel specified the institutions which were to pick the winners of his prizes. Physics and chemistry laureates were to be chosen by committees of Sweden's Royal Academy of Science, medicine laureates by the Royal Caroline Institute of Medicine. These procedures have never changed. Each year nearly 2,000 requests for recommendations are sent out: to past laureates, university professors all over the world, including Sweden, and other authorities. Hundreds of recommendations come in. The final choices are made in secret meetings, after lengthy investigations and discussions. Finally, about a month before the award ceremony, matter-of-fact telegrams inform the new Nobel laureates of their selection. In recent years, partly because scientists have tended more and more toward group research, and partly because of breakthroughs in highly specialized fields, the prize money has often been shared by two or three laureates in each category.

N:o 79. År 1897 den 5 Februari uppvisat vid Testamentsdelen inför Stockholms Rådstufvurätts Sjette Afdelning. Betygar ex officio Jacob Lundell

Lösen En krona ant. å prot.

Testament

Jag undertecknad Alfred Bernhard Nobel förklarar härmed efter moget betänkande min yttersta vilja i afseende å den egendom jag vid min död kan efterlemna vara följande:

Mina Brorssöner Hjalmar och Ludvig Nobel, söner af min Broder Robert Nobel, erhålla hvardera en summa af Två Hundra Tusen Kronor;

Min Brorson Emmanuel Nobel erhåller Tre Hundra Tusen och min Brorsdotter Mina Nobel Ett Hundra Tusen kronor;

Min Broder Robert Nobels döttrar Ingeborg och Tyra erhålla hvardera Ett Hundra Tusen kronor;

Fröken Olga Boettger, för närvarande boende hos Fru Brand, 10 Rue St Florentin i Paris, erhåller Ett Hundra Tusen Francs;

Fru Sofie Kapy von Kapivar, hvars adresse är känd af Anglo-Oesterreichische Bank i Wien är berättigad till en lifränta af 6000 Florins Ö.W. som betalas henne af sagde Bank och hvarföre jag i denna Bank deponerat 150,000 Fl. i Ungerska Statspapper.

Herr Alarik Liedbeck, boende 26 Sturegatan Stockholm, erhåller Ett Hundra Tusen Kronor;

Fröken Elise Antun, boende 32 Rue de Lubeck Paris, är berättigad till en lifränta af Två Tusen Fem Hundra Francs. Dessutom innestår hos mig för närvarande Fyratio åtta Tusen Francs henne tillhörigt Kapital som äfven skall till henne återbetalas.

Herr Alfred Hammond, Waterford, Texas, United States, erhåller Tio Tusen Dollars;

Fröknarna Emmy Winkelmann och Marie Win-

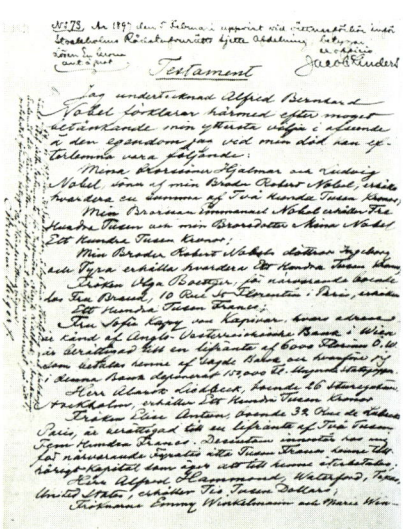

NOBEL'S LAST WILL AND TESTAMENT
This was the final will of Alfred Nobel. He composed it without the aid of a lawyer. Sideways in the left margin and at the top are legal certifications. Some of Nobel's relatives briefly attempted to contest Nobel's wishes, but gave up.

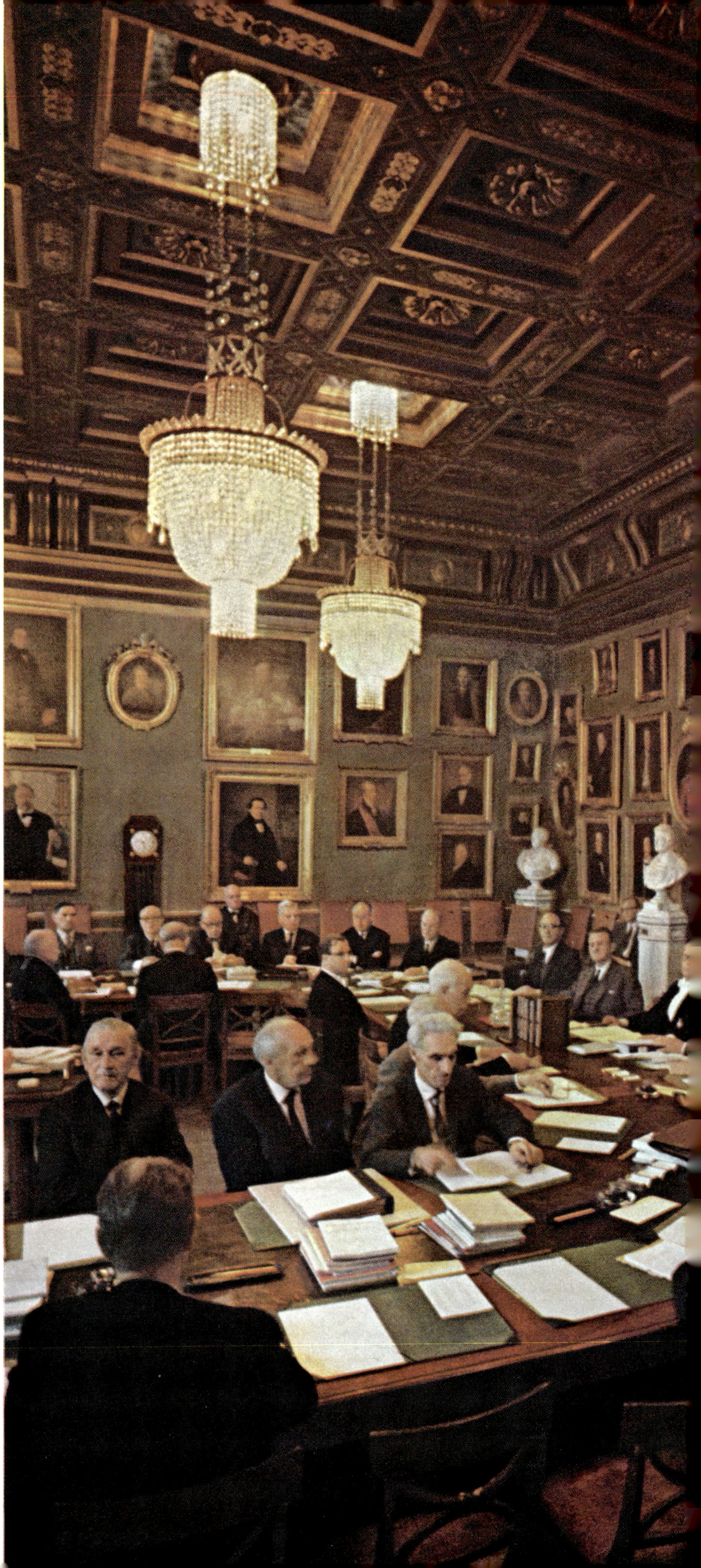

A LIBRARY OF RECOMMENDATIONS
Arne Ölander, secretary of the Nobel committees for chemistry and physics, looks through one of the bound volumes of recommendations (blue, physics; red, chemistry). All proposals received are included, even crank letters from people claiming to have a cure for all known diseases.

THE MEDICINE COMMITTEE
The Nobel Committee for Medicine of the Royal Caroline Institute *(below)* convenes in a room decorated with a portrait of Alfred Nobel. To ensure a wide range of judgment, the members of the committee are chosen from as many different fields of medical research as possible.

```
WUA002 (O CDV156 WUH232 SWB0454) 60 PD INTL FR CD STOCKHOLM VIA
RCA 5 1245
PROFESSOR M SG  GOEPPERT - MAYER
2345 VIASIENNA LAJOLLA (CALIF)
ROYAL ACADMY OF SCIENCE TODAY      AWARDED YOU AND JOPSEN JOINTLY
ONE HALF OF THE 1963 NOBEL PRIZE FOR PHYSICS FOR YOUR DISCOVERIES
CONCERNING NUCLEAR SHELL STRUCTURE THE OTHER HALF SENT TO WIGNER
TOR THEORY FOR NUCLEUS AND ELEMENATARY  PARTICLES ESPECIALLY
DISCOVERY AND APPLICATION OF FUNDAMAMENTAL SYMMETRY PRINCIPLES
LETTER FOLLOWS
       RUDBERG PERMANENT SECRETARY
```

THE ROYAL ACADEMY OF SCIENCE
At a regular meeting of the Royal Academy of Science, members discuss finance matters of income and properties owned by the Nobel Foundation. Surrounding the men are portraits and busts of former members of the Academy.

THE TELEGRAM
This wire—with a co-winner's name (Jensen) misspelled—brought first word to Maria Goeppert Mayer of her selection as physics laureate. She recalls that, as she put it, she "did the obvious thing and opened a bottle of champagne."

The Great Moment

At 4:30 in the afternoon of December 10, the anniversary of the death of Alfred Nobel, the laureates, each accompanied by a sponsor who is a member of the Nobel Foundation, take their positions on the stage of the Concert Hall in Stockholm. Earlier in the day, to empty seats, they had carefully rehearsed the ceremony. Now every seat is filled. The hall is hushed. Before each award is made, the work that earned it is briefly summarized by a member of the Foundation—in Swedish. Then the winner is asked—in his own language—to come forward. He walks down the steps to stand before the King of Sweden. The King hands the laureate a diploma and a medal, and speaks a few quiet words of congratulation. The audience offers a standing ovation.

THE SETTING AND THE AWARD
In the flower-bedecked Concert Hall *(above)*, the laureates sit on the platform as 2,000 silent spectators await the presentations. In the picture at the right, King Gustaf Adolf *(back to camera)* is presenting a medal and diploma to physicist Eugene P. Wigner. Co-winner Professor Maria Goeppert Mayer, the only woman in addition to Marie Curie to receive the award for physics, is waiting her turn. J. Hans D. Jensen is descending the steps from the stage.

A BANQUET FOR LAUREATES
By candle and torchlight, nearly 800 guests enjoy a two-hour banquet served in Stockholm's City Hall. Seated at the long head table *(right)* are members of the Royal Family, the laureates and their families, and other public figures.

A PANOPLY OF BANNERS
As the Royal Family watches from the balcony, white-capped Swedish students welcome the laureates to the ball. The banners they carry bear the insignia of the University of Stockholm and a number of institutions of higher learning.

QUEEN AND PHYSIOLOGIST
At the head table, Sweden's Queen Louise talks earnestly with Australian Sir John Carew Eccles, who together with Alan Hodgkin and Andrew Huxley was awarded the 1963 prize for medicine. Many laureates have been surprised to find that both the Queen and the King are well informed on the latest developments in all of the sciences, but especially in medicine.

A Glittering Finale

For most winners of the Nobel Prize, the trip to Sweden takes on an air of the fantastic. For their entire stay, they are guests of the Nobel Foundation. At the Grand Hotel in Stockholm they stay in the Nobel suite; chauffeur-driven limousines are put at their disposal. Pomp and majesty attend every function.

The award ceremony itself is so formal that even photographers wear white tie and tails. The menu usually offers caviar, venison, fish cooked in champagne, and fine wines and liqueurs. At the ball, although it, too, is a glittering affair, the laureates have a chance to loosen up—a few of the winners have been known to see the night out with the students, who leave only when the band does.

After the splendors of award day, it seems anticlimactic when, the next day, the laureates drop by the Nobel Foundation to pick up their checks.

WILHELM CONRAD RÖNTGEN, PHYSICS, 1901

IVAN PETROVIČ PAVLOV, MEDICINE, 1904

MARIE SKLODOWSKA CURIE, CHEMISTRY, 1911

WILLIAM LAWRENCE BRAGG, PHYSICS, 1915

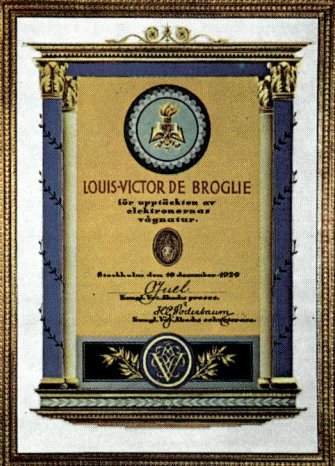

PRINCE LOUIS-VICTOR DE BROGLIE, PHYSICS, 1929

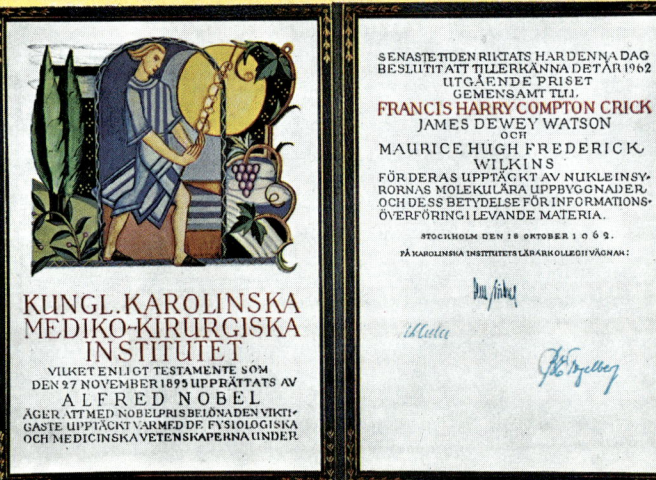

FRANCIS HARRY COMPTON CRICK, MEDICINE, 1962

THE DIPLOMAS

The Nobel diplomas are individually designed for each winner in a style resembling medieval illuminated manuscripts. In design motifs they range from stylized Greek temples (for Louis-Victor de Broglie) to involved symbolism (for Francis H. C. Crick). In some cases, the decoration is closely related to the work of the laureate: a gas-discharge tube for Röntgen *(top left)*.

Testaments in Gold and Parchment

The Nobel medals, first minted in 1902 (1901 winners received theirs a year late), are of 23-karat gold, about two and one half inches across, and weigh nearly half a pound. They were designed by Swedish sculptor Erik Lindberg, and his bas-relief of Nobel in profile is considered to be one of the finest likenesses ever made of the founder. The scene on the obverse of the physics and chemistry medals depicts the goddess Isis, the Earth Mother, holding a horn of plenty, and the genius of Science lifting the veil from her face.

No two diplomas are ever alike—with one exception: the diploma for William Lawrence Bragg (*middle row, opposite*) is identical in design with the one given his father, Sir William Henry Bragg. The two, who were in 1915 jointly awarded the prize for physics, were the only father-son team ever to share a Nobel Prize.

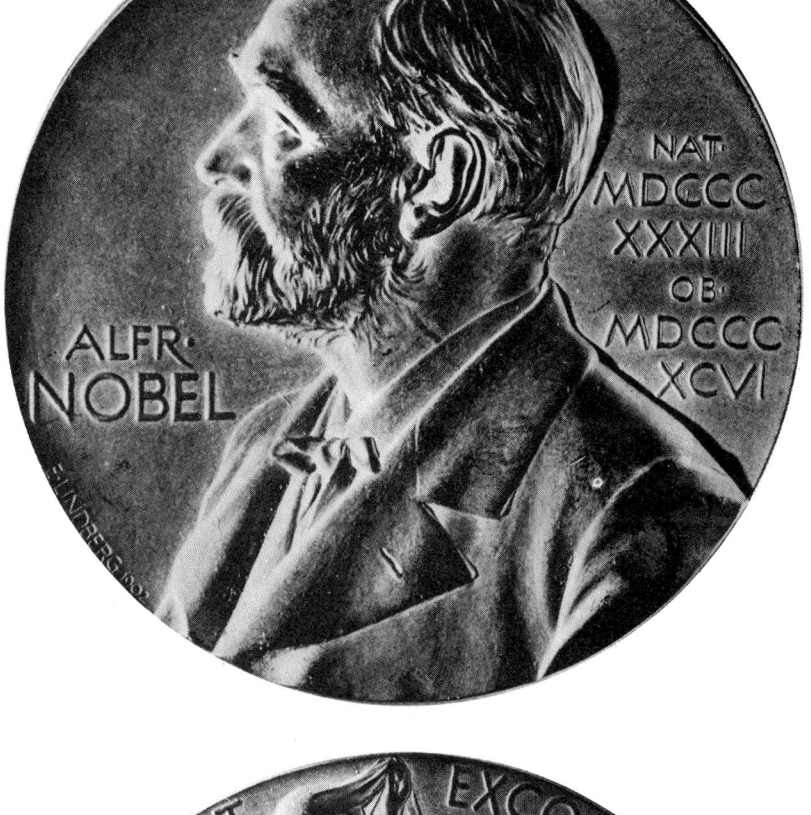

THE MEDAL
The front of the medal is devoted to Alfred Nobel. The dates of his birth and death are given in Roman numerals. On the obverse are the winner's name, the year he won (here, Albert Einstein, 1921), and several inscriptions in Latin: "natural science," abbreviations of the name of the Royal Academy of Science of Sweden, and a statement paraphrased from Virgil's *Aeneid:* "How good it is that man's life should be enriched by the arts he has invented."

"A FRANTIC FEW DAYS"
F.H.C. Crick, co-winner of the 1962 medicine prize, takes the dance floor with his daughter Gabrielle. Mrs. Crick *(left)*, who said some prize money would be spent on a yacht, described the family's stay in Stockholm as a "frantic few days of champagne, dancing and gaiety."

The Joy of Winning

How does it feel to win the Nobel Prize? Sir John Cockcroft, physics laureate in 1951, said he felt "transported into a magical world by the genie of Alfred Nobel." Peter Medawar, winner of the medicine prize in 1960, felt as though he had "suddenly been transported from an ivory tower to a golden palace."

There is a bread-and-butter aspect of the joy of winning, too—the prize money. The money allows some recipients to carry their research forward: in the case of Marie and Pierre Curie, for example, who together won half the 1903 award for physics, the stipend allowed Pierre to give up teaching and concentrate on research.

At least one laureate emulated Alfred Nobel: Lord Rayleigh, winner of the physics award of 1904, donated nearly half the prize money to the internationally famous Cavendish Laboratory, in Cambridge, England, where he had discovered the element argon, for which he won the prize.

"A HAPPY SURPRISE"
Grinning at an unseen jokester, Alan Hodgkin of Britain plays his role in the rehearsal for the award ceremony. Hodgkin shared the 1963 medicine prize for his research on nerve cells. His understated reaction to the news of his selection: "It certainly was a happy surprise."

HIGH-STEPPING PHYSICIST
Chen Ning Yang and his wife do some fancy steps at the 1957 ball. Yang, then 35, and his collaborator, Tsung Dao Lee, 31, shared that year's award for physics for their work in the investigation of the parity laws *(page 8)*.

APPENDIX

A Gallery of Nobel Laureates

To a great degree, the Nobel awards—first presented in 1901—chronicle the milestone developments in physics, chemistry and medicine that have marked 20th Century scientific history. The photographs on this and the following pages include all the science laureates, arranged in specific categories such as atomic physics, optics and physical chemistry. Within these categories, the winners are listed chronologically according to the year of the prize.

Through this arrangement, progress in many fields can be traced from the first experimental work of the pioneer researcher to the infinitely complicated studies of modern scientists. Under Leading Theorists, for example, it is possible to follow quantum theory from its conception, for which Max Planck received the physics award in 1918.

An examination of the Nobel lists discloses some surprises. The work for which a prize was awarded was not always a laureate's most important contribution. Einstein, for instance, won the prize for physics in 1921 not for his famous theory of relativity, which he had published 16 years earlier, but for his less well-known work on the photoelectric effect. Five awards have gone to members of a single family. In 1903 Marie Curie and her husband Pierre shared the prize for physics with Henri Becquerel *(page 190)*. In 1911 Marie Curie was again a laureate, this time in chemistry. In 1935 her daughter and son-in-law, Irène Joliot-Curie and her husband Frédéric Joliot, shared the prize for chemistry.

LEADING THEORISTS
Max Planck, German, 1918: for his hypothesis that all radiation was emitted in units called quanta.
Albert Einstein, German, 1921: for services to physics and the law of the photoelectric effect.
Niels Bohr, Danish, 1922: for his atomic theory that laid the groundwork for later atomic research.
Louis de Broglie, French, 1929: for discovering the wave nature of electrons.
Werner Heisenberg, German, 1932: for the theory that began modern quantum mechanics.
Erwin Schrödinger, Austrian, and **Paul Dirac**, English, 1933: for new atomic theory, including wave mechanics and prediction of the positron.
Wolfgang Pauli, Austrian, 1945: for the principle governing the arrangement of electrons in atomic orbits.
Max Born, English, 1954: for his statistical interpretation of the quantum mechanical wave function.

ATOMIC PHYSICS
Wilhelm Röntgen, German, 1901: for his discovery of the X-ray.
Hendrik Lorentz and **Pieter Zeeman**, Dutch, 1902: for their discovery of the influence of magnetism on beams of light.
Philipp von Lenard, German, 1905: for his work with cathode rays.
Joseph Thomson, English, 1906: for his discovery of the electron and measurements of the ratio of its charge and mass.
Charles Barkla, English, 1917: for discovering characteristic X-rays of each element.
Johannes Stark, German, 1919: for finding the Doppler effect in canal rays and for splitting the lines of the spectrum.
Robert Millikan, American, 1923: for work on the photoelectric effect and the charge of the electron.
Karl Siegbahn, Swedish, 1924: for discoveries and research in X-ray spectroscopy.
James Franck and **Gustav Hertz**, Germans, 1925: for experiments testing Planck's quantum theory and laws for electron-atom collisions.
Arthur Compton, American, 1927: for his discovery of the Compton effect in X-ray research.
Clinton Davisson, American, and **George Thomson**, English, 1937: for experimental discovery of the diffraction of electrons by crystals.
Willis Lamb and **Polykarp Kusch**, Americans, 1955:

Lamb for work on the fine structure of the hydrogen spectrum, and Kusch for measuring the magnetic moment of the electron.

NUCLEAR PHYSICS

Henri Becquerel, Pierre and **Marie Curie**, French, 1903: Becquerel for discovering radioactivity in uranium, and the Curies for discovering other radioactive elements.

Charles Wilson, English, 1927: for inventing the cloud chamber for observing nuclear particles.

James Chadwick, English, 1935: for the discovery of the neutron.

Victor Hess, Austrian, and **Carl Anderson**, American, 1936: Hess for discovering cosmic radiation, and Anderson for his discovery of the positron.

Enrico Fermi, Italian, 1938: for identifying new elements and discovering nuclear reactions by his methods of nuclear irradiation and bombardment.

Ernest Lawrence, American, 1939: for his invention of the cyclotron and some experiments on radioactive elements made with it.

Otto Stern, American, 1943: for the molecular ray method and measurement of the proton magnetic moment by this method.

Otto Hahn, German, 1944: for discovering the fission of heavy atomic nuclei.

Isidor Rabi, American, 1944: for the resonance method of measuring nuclear magnetic properties.

Patrick Blackett, English, 1948: for his method of photographing the positron.

Hideki Yukawa, Japanese, 1949: for predicting mesons as the field of the nuclear force.

Cecil Powell, English, 1950: for methods of photographing nuclear processes, and confirming the existence of mesons.

Sir John Cockcroft, English, and **Ernest Walton**, Irish, 1951: for transmutation of atomic nuclei with artificially accelerated atomic particles.

Felix Bloch and **Edward Purcell**, Americans, 1952: for new methods of nuclear magnetic measurements and subsequent discoveries.

Walther Bothe, German, 1954: for a significant improvement of the Geiger-Müller tube.

Chen Ning Yang and **Tsung Dao Lee**, Chinese, 1957: for their investigation of parity laws and the implications for elementary particles.

Pavel Čerenkov, Il'ja Frank and **Igor' Tamm**, Russians, 1958: for discovering and interpreting the Čerenkov effect—that fast electrons emit a blue glow when traveling through a transparent substance such as water.

Emilio Segrè and **Owen Chamberlain**, Americans, 1959: for their discovery of the antiproton.

Donald Glaser, American, 1960: for the invention of the bubble chamber for observing the paths of nuclear particles.

Robert Hofstadter, American, and **Rudolf Mössbauer**, German, 1961: Hofstadter for studies of electron-scattering in atomic nuclei, and Mössbauer for discovering the Mössbauer effect—a method for producing gamma rays having very precise wavelengths.

Eugene Wigner, American, **Maria Goeppert Mayer** and **Hans Jensen**, Germans, 1963: Wigner for discovering and applying principles of fundamental particle symmetry, and Mayer and Jensen for their discoveries in nuclear shell structure.

SOLID-STATE PHYSICS

Max von Laue, German, 1914: for his discovery that X-rays can be diffracted into meaningful patterns by crystals, which enabled the wavelengths of X-rays to be determined and the atomic structure of crystals to be revealed.

Sir William H. Bragg and **William L. Bragg**, English, 1915: for further studies of crystal structure by means of X-rays and for the mathematical equation that explains the resulting diffraction pattern.

Charles Guillaume, French, 1920: for the production of nickel-steel alloys whose properties made them suitable for use in precision instruments for measuring length and mass.

Owen Richardson, English, 1928: for his work on the phenomenon of thermionic emission—or the fact that air next to a glowing metal body conducts electricity—and especially for the formulation of the law named after him.

William Shockley, John Bardeen and **Walter Brattain**, Americans, 1956: for their research on semiconductors such as silicon and germanium, and their discovery of the transistor, which replaced the vacuum tube as a power amplifier in the field of electronics.

OPTICS

Albert Michelson, American, 1907: for development of precise tools for measuring properties of light.

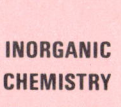

Sir Chandrasekhara Raman, Indian, 1930: for work on the component wavelengths of a ray of light.
Frits Zernike, Dutch, 1953: for inventing the phase-contrast microscope.

THERMODYNAMICS

Johannes van der Waals, Dutch, 1910: for investigations of pressure, mass and temperature in gases and liquids.
Wilhelm Wien, German, 1911: for discoveries pertaining to the radiation of heat.
Heike Kamerlingh-Onnes, Dutch, 1913: for studies of the properties of matter at very low temperatures.
Walther Nernst, German, 1920: for a new heat theorem and other basic work in thermochemistry.
William Giauque, American, 1949: for studies on the behavior of substances at extremely low temperatures.
Lev Landau, Russian, 1962: for theories about condensed matter, especially liquid helium.

APPLIED PHYSICS

Gabriel Lippmann, French, 1908: for his method of reproducing colors photographically by means of refracted light.
Guglielmo Marconi, Italian, and **Carl Braun**, German, 1909: for contributions to the development of wireless telegraphy.
Nils Dalén, Swedish, 1912: for inventing automatic regulators used in gas illumination.
Sir Edward Appleton, English, 1947: for investigations of the physics of the upper atmosphere.

PHYSICAL CHEMISTRY

Jacobus van't Hoff, Dutch, 1901: for his discovery of laws of chemical reactions and of osmotic pressure.
Svante Arrhenius, Swedish, 1903: for explaining how elements in solution are separated by electrolysis.
Henri Moissan, French, 1906: for isolating fluorine.
Ernest Rutherford, English, 1908: for his investigations of the chemistry of radioactive substances.
Wilhelm Ostwald, German, 1909: for discovering principles of chemical equilibriums and rates of reaction.
Francis Aston, English, 1922: for discovering isotopes of many nonradioactive elements.
Richard Zsigmondy, German, 1925: for demonstrating the heterogeneous nature of colloidal solutions.
Jean Perrin, French, 1926: for work on laws of the distribution of particles in matter.
The Svedberg, Swedish, 1926: for work on colloidal solutions.
Carl Bosch and **Friedrich Bergius**, Germans, 1931: for the development of chemical high-pressure methods.
Irving Langmuir, American, 1932: for discoveries in surface chemistry.
Petrus Debye, Dutch, 1936: for contributions to knowledge of molecular structure.
Percy Bridgman, American, 1946: for discoveries in the field of high-pressure physics.
Linus Pauling, American, 1954: for research into the nature of the chemical bond.
Sir Cyril Hinshelwood, English, and **Nikolaj Semenov**, Russian, 1956: for studies in the mechanism of chemical reactions.
Jaroslav Heyrovský, Czech, 1959: for developing polarographic methods of chemical analysis.
Willard Libby, American, 1960: for developing the carbon-14 dating method.

INORGANIC CHEMISTRY

Sir William Ramsay, English, 1904: for discovering inert gases in the air and placing them in the periodic table.
John Lord Rayleigh, English, 1904: for investigating the density of gases and discovering argon.
Marie Curie, French, 1911: for the discovery of the elements radium and polonium, and for the isolation of radium and the study of its nature and compounds.
Alfred Werner, Swiss, 1913: for investigating the linkage of atoms in molecules, opening new fields of study in inorganic chemistry.
Theodore Richards, American, 1914: for determining the atomic weights of many chemical elements.
Fritz Haber, German, 1918: for the synthesis of ammonia from its elements.
Frederick Soddy, English, 1921: for work in the chemistry of radioactive substances and the origin and nature of isotopes.
Harold Urey, American, 1934: for discovering heavy hydrogen.
Frédéric Joliot and **Irène Joliot-Curie**, French, 1935: for synthesizing new radioactive elements.
Edwin McMillan and **Glenn Seaborg**, Americans, 1951: for discovering new elements heavier than uranium, called transuranic elements.

ORGANIC CHEMISTRY

Emil Fischer, German, 1902: for producing sugars synthetically.

Johann von Baeyer, German, 1905: for work that led to the development of organic dyes and medicines.

Otto Wallach, German, 1910: for explaining the construction of alicyclic compounds (volatile oils).

Victor Grignard and **Paul Sabatier**, French, 1912: Grignard for synthesizing, and Sabatier for hydrogenating, organic compounds.

Fritz Pregl, Austrian, 1923: for developing a method of microanalysis of organic substances.

Leopold Ružicka, Swiss, 1939: for discoveries of the structure of certain odorous substances, such as polymethylenes and higher terpenes.

Otto Diels and **Kurt Alder**, Germans, 1950: for developing a method of synthesis useful in the plastics industry.

Hermann Staudinger, German, 1953: for discoveries of the behavior of large molecules, which led to developments in plastics and synthetic fibers.

Karl Ziegler, German, and **Giulio Natta**, Italian, 1963: for their discoveries in the chemistry and technological application of high polymers as used in plastics, films and synthetic fibers.

BIOCHEMISTRY

Eduard Buchner, German, 1907: for his discovery of cell-free fermentation.

Albrecht Kossel, German, 1910: for pioneer work on the chemistry of proteins and nucleic substances.

Richard Willstätter, German, 1915: for research on chlorophyll and other plant pigments.

Frederick Banting and **John Macleod**, Canadians, 1923: for the discovery of insulin.

Heinrich Wieland, German, 1927: for investigating the chemical constitution of bile acids.

Adolf Windaus, German, 1928: for research on the sterols and their connection with vitamins.

Christiaan Eijkman, Dutch, and **Sir Frederick Hopkins**, English, 1929: for the discoveries, respectively, of antineuritic vitamins and of growth-stimulating vitamins.

Arthur Harden, English, and **Hans von Euler-Chelpin**, Swedish, 1929: for investigations into sugar fermentation and fermentative enzymes.

Hans Fischer, German, 1930: for research in the chemical structure of hemin and chlorophyll, and especially for the synthesis of hemin.

Otto Warburg, German, 1931: for his discovery of the nature and workings of the respiratory enzyme.

Walter Haworth, English, and **Paul Karrer**, Swiss, 1937: Haworth for investigations of carbohydrates and vitamin C, Karrer for studies of carotenoids, flavin, vitamins A and B$_2$.

Albert Szent-Györgyi von Nagyrapolt, Hungarian, 1937: for discoveries in biological combustion processes.

Richard Kuhn, German, 1938: for work on the chemistry of carotenoids and vitamins.

Adolf Butenandt, German, 1939: for his work on the chemistry of the sex hormones.

Henrik Dam, Danish, and **Edward Doisy**, American, 1943: for discovery of vitamin K and its chemical composition.

George de Hevesy, Hungarian, 1943: for developing radioactive isotopes as laboratory tracers in research on chemical processes.

Artturi Virtanen, Finnish, 1945: for inventions in agricultural and nutrition chemistry, especially for his method of fodder preservation.

James Sumner, **John Northrop** and **Wendell Stanley**, Americans, 1946: Sumner for discovering that enzymes can be crystallized, Northrop and Stanley for isolating enzymes and virus proteins in pure form.

Carl and **Gerty Cori**, Americans, and **Bernardo Houssay**, Argentinian, 1947: the Coris for discovering how glycogen is catalytically converted, and Houssay for analyzing how the anterior lobe of the pituitary gland affects the metabolism of sugar.

Sir Robert Robinson, English, 1947: for biological research on plant products, particularly alkaloids.

Arne Tiselius, Swedish, 1948: for discoveries in the complex chemistry of the serum proteins.

Edward Kendall, **Philip Hench**, Americans, and **Tadeus Reichstein**, Swiss, 1950: for discoveries involving hormones of the cortex of the adrenal gland during which they isolated cortisone.

Archer Martin and **Richard Synge**, English, 1952: for inventing partition chromatography, a method of separating complex biochemical compounds used in finding new antibiotics.

Hans Krebs, English and **Fritz Lipmann**, American, 1953: for discoveries, respectively, of the citric

acid cycle and of coenzyme A, important in intermediary metabolism.
Axel Theorell, Swedish, 1955: for discoveries involving enzymes that turn food into energy.
Vincent du Vigneaud, American, 1955: for work done on the chemistry of biologically important sulphur compounds.
Daniel Bovet, Italian, 1957: for discoveries of synthetic compounds useful in the treatment of allergies.
Sir Alexander Todd, English, 1957: for work done in the biochemistry of metabolism.
Frederick Sanger, English, 1958: for his work in analyzing the structure of proteins, especially insulin.
Severo Ochoa and **Arthur Kornberg,** Americans, 1959: for discoveries in the biochemistry of cells.
Melvin Calvin, American, 1961: for studies of how plants assimilate carbon dioxide from air.
Francis Crick and **Maurice Wilkins,** English, and **James Watson,** American, 1962: for discoveries in the chemical structure of genes, and how they regulate the formation of new living tissue.
Max Perutz and **John Kendrew,** English, 1962: for analysis of globular proteins.

GENETICS
Thomas Morgan, American, 1933: for his discoveries of the ways in which chromosomes govern heredity.
Hermann Muller, American, 1946: for discovering that genetic mutations can be induced by X-ray irradiation.
George Beadle, Edward Tatum and **Joshua Lederberg,** Americans, 1958: Beadle and Tatum for the discovery that genes are the regulators of definite chemical events, Lederberg for new theories in the area of genetic mechanics—how the genes of bacteria are organized.

PHYSIOLOGY
Ivan Pavlov, Russian, 1904: for pioneer work in the study of the digestive system.
Allvar Gullstrand, Swedish, 1911: for research on the way the eye registers light and forms images.
Schack Krogh, Danish, 1920: for his discovery of the way the capillaries regulate blood circulation by altering their dimensions.
Archibald Hill, English, and **Otto Meyerhof,** German, 1922: for discoveries showing that heat is produced and oxygen consumed in muscles.
Willem Einthoven, Dutch, 1924: for inventing the electrocardiogram.
Karl Landsteiner, Austrian, 1930: for the discovery of human blood groups.
Hans Spemann, German, 1935: for discoveries in embryology involving development of the central nervous system.
Corneille Heymans, Belgian, 1938: for discovery of the roles of the sinus and aorta in respiration.
Georg von Békésy, American, 1961: for discoveries of the workings of the cochlea in the inner ear.

NEUROPHYSIOLOGY
Camillo Golgi, Italian, and **Santiago Ramon y Cajal,** Spanish, 1906: for their studies of the structure of nerve cells, including a new method of color preparation.
Robert Bárány, Austrian, 1914: for work on the physiology and pathology of the vestibular apparatus of the inner ear, which controls the sense of balance.
Sir Charles Sherrington and **Edgar Adrian,** English, 1932: for their discoveries regarding the functions of neurons, or nerve cells, as transmitters of nervous energy along circuits in the nervous system.
Sir Henry Dale, English, and **Otto Loewi,** Austrian, 1936: for their explanation of the chemical transmission of nerve impulses by their discovery of acetylcholine, the substance given off at the tips of nerve fibers.
Joseph Erlanger and **Herbert Gasser,** Americans, 1944: for their discoveries relating to different functions of single nerve fibers, and their recording of the electrical impulses involved in nerve impulses on a cathode-ray oscillograph.
Walter Hess, Swiss, and **Antonio Moniz,** Portuguese, 1949: Hess for his discovery of the organization of the midbrain as coordinator of activities of the internal organs, and Moniz for his discovery of the value of the frontal lobotomy in certain psychoses.
Sir John Eccles, Australian, and **Alan Hodgkin** and **Andrew Huxley,** English, 1963: for discoveries concerning the sodium and potassium ion mechanisms involved in the transmission of nerve impulses across the membranes of nerve cells.

MEDICINE

Emil von Behring, German, 1901: for his work on serum therapy, especially a serum to combat diphtheria.

Sir Ronald Ross, English, 1902: for his discovery that malaria is spread by the anopheles mosquito, and his researches in prevention.

Niels Finsen, Danish, 1903: for his research on the use of light radiation as a cure for disease, especially tuberculosis.

Robert Koch, German, 1905: for successfully isolating the bacillus causing tuberculosis, and for proposing a cure, tuberculin (later found to be unsuccessful).

Charles Laveran, French, 1907: for his work on the role played by protozoa in causing disease, especially malaria.

Paul Ehrlich, German, and **Il'ja Mečnikov,** Russian, 1908: for their work on immunity—Ehrlich for his discovery of how antibodies are formed, and Mečnikov for his discovery of white corpuscles.

Emil Kocher, Swiss, 1909: for his research into the functions and for developing surgical treatment of the thyroid gland.

Alexis Carrel, American, 1912: for his work on transplantation of blood vessels and organs, laying the foundation for a new field of surgical cures.

Charles Richet, French, 1913: for his work on anaphylaxis, a state of increased susceptibility to poisons following an injection.

Jules Bordet, Belgian, 1919: for discoveries relating to immunity.

Johannes Fibiger, Danish, 1926: for his discovery of the cancer, spiroptera carcinoma.

Julius Wagner-Jauregg, Austrian, 1927: for the development of malaria inoculation as a cure for general paresis.

Charles Nicolle, French, 1928: for his work on typhus, including his discovery of how the disease is spread.

George Whipple, George Minot and **William Murphy,** Americans, 1934: for the discovery of liver therapy in cases of anemia.

Gerhard Domagk, German, 1939: for the discovery of the antibacterial effects of the drug prontosil, the forerunner of the sulfa drugs.

Sir Alexander Fleming, Ernst Chain and **Sir Howard Florey,** English, 1945: for the discovery of penicillin, a powerful antibiotic attacking many infectious diseases.

Paul Muller, Swiss, 1948: for his discovery of the efficiency of DDT as an insecticide, which proved to be especially useful in the extermination of the malaria mosquito.

Max Theiler, South African, 1951: for his research into the nature of, and cures for, the virus disease, yellow fever.

Selman Waksman, American, 1952: for his discovery of streptomycin, the first antibiotic effective against tuberculosis.

John Enders, Thomas Weller and **Frederick Robbins,** Americans, 1954: for their discovery that poliomyelitis viruses can grow in cultures of various kinds of tissue.

André Cournand, Dickinson Richards Jr., Americans, and **Werner Forssmann,** German, 1956: for their discoveries of heart catheterization and pathological changes in the circulatory system.

Sir F. Macfarlane Burnet, Australian, and **Peter Medawar,** English, 1960: for their discovery of acquired immunological tolerance in transplants.

ALPHABETICAL LIST OF NOBEL PRIZE WINNERS

A
Adrian, Edgar Douglas: Medicine, 1932. *P. 193*
Alder, Kurt: Chemistry, 1950. *P. 192*
Anderson, Carl David: Physics, 1936. *P. 190*
Appleton, Sir Edward Victor: Physics, 1947. *P. 191*
Arrhenius, Svante August: Chemistry, 1903. *P. 191*
Aston, Francis William: Chemistry, 1922. *P. 191*

B
Baeyer, Johann Friedrich Wilhelm Adolf von: Chemistry, 1905. *P. 192*
Banting, Frederick Grant: Medicine, 1923. *P. 192*
Bárány, Robert: Medicine, 1914. *P. 193*
Bardeen, John: Physics, 1956. *P. 190*
Barkla, Charles Glover: Physics, 1917. *P. 189*
Beadle, George Wells: Medicine, 1958. *P. 193*
Becquerel, Antoine Henri: Physics, 1903. *P. 190*

Behring, Emil Adolf von: Medicine, 1901. *P. 194*
Békésy, Georg von: Medicine, 1961. *P. 193*
Bergius, Friedrich: Chemistry, 1931. *P. 191*
Blackett, Patrick Maynard Stuart: Physics, 1948. *P. 190*
Bloch, Felix: Physics, 1952. *P. 190*
Bohr, Niels: Physics, 1922. *P. 189*
Bordet, Jules: Medicine, 1919. *P. 194*
Born, Max: Physics, 1954. *P. 189*
Bosch, Carl: Chemistry, 1931. *P. 191*
Bothe, Walther: Physics, 1954. *P. 190*
Bovet, Daniel: Medicine, 1957. *P. 193*
Bragg, Sir William Henry: Physics, 1915. *P. 190*
Bragg, William Lawrence: Physics, 1915. *P. 190*
Brattain, Walter Houser: Physics, 1956. *P. 190*
Braun, Carl Ferdinand: Physics, 1909. *P. 191*
Bridgman, Percy Williams: Physics, 1946. *P. 191*
Broglie, Prince Louis-Victor de: Physics, 1929. *P. 189*

Buchner, Eduard: Chemistry, 1907. *P. 192*
Burnet, Sir F. Macfarlane: Medicine, 1960. *P. 194*
Butenandt, Adolf Friedrich Johann: Chemistry, 1939. *P. 192*

C
Calvin, Melvin: Chemistry, 1961. *P. 193*
Carrel, Alexis: Medicine, 1912. *P. 194*
Čerenkov, Pavel Aleksejevič: Physics, 1958. *P. 190*
Chadwick, James: Physics, 1935. *P. 190*
Chain, Ernst Boris: Medicine, 1945. *P. 194*
Chamberlain, Owen: Physics, 1959. *P. 190*
Cockcroft, Sir John Douglas: Physics, 1951. *P. 190*
Compton, Arthur Holly: Physics, 1927. *P. 189*
Cori, Carl Ferdinand: Medicine, 1947. *P. 192*
Cori, Gerty Theresa: Medicine, 1947. *P. 192*
Cournand, André Frédéric: Medicine, 1956. *P. 194*

Crick, Francis Harry Compton: Medicine, 1962. *P. 193*
Curie, Marie: Physics, 1903. *P. 190*
Curie, Marie: Chemistry, 1911. *P. 191*
Curie, Pierre: Physics, 1903. *P. 190*

D

Dale, Sir Henry Hallett: Medicine, 1936. *P. 193*
Dalén, Nils Gustaf: Physics, 1912. *P. 191*
Dam, Henrik Carl Peter: Medicine, 1943. *P. 192*
Davisson, Clinton Joseph: Physics, 1937. *P. 189*
Debye, Petrus Josephus Wilhelmus: Chemistry, 1936. *P. 191*
Diels, Otto Paul Hermann: Chemistry, 1950. *P. 192*
Dirac, Paul Adrien Maurice: Physics, 1933. *P. 189*
Doisy, Edward Adelbert: Medicine, 1943. *P. 192*
Domagk, Gerhard: Medicine, 1939. *P. 194*

E

Eccles, Sir John Carew: Medicine, 1963. *P. 193*
Ehrlich, Paul: Medicine, 1908. *P. 194*
Eijkman, Christiaan: Medicine, 1929. *P. 192*
Einstein, Albert: Physics, 1921. *P. 189*
Einthoven, Willem: Medicine, 1924. *P. 193*
Enders, John Franklin: Medicine, 1954. *P. 194*
Erlanger, Joseph: Medicine, 1944. *P. 193*
Euler-Chelpin, Hans Karl August Simon von: Chemistry, 1929. *P. 192*

F

Fermi, Enrico: Physics, 1938. *P. 190*
Fibiger, Johannes Andreas Grib: Medicine, 1926. *P. 194*
Finsen, Niels Ryberg: Medicine, 1903. *P. 194*
Fischer, Hans: Chemistry, 1930. *P. 192*
Fischer, Hermann Emil: Chemistry, 1902. *P. 192*
Fleming, Sir Alexander: Medicine, 1945. *P. 194*
Florey, Sir Howard Walter: Medicine, 1945. *P. 194*
Forssmann, Werner: Medicine, 1956. *P. 194*
Franck, James: Physics, 1925. *P. 189*
Frank, Il'ja Michajlovič: Physics, 1958. *P. 190*

G

Gasser, Herbert Spencer: Medicine, 1944. *P. 193*
Giauque, William Francis: Chemistry, 1949. *P. 191*
Glaser, Donald Arthur: Physics, 1960. *P. 190*
Golgi, Camillo: Medicine, 1906. *P. 193*
Grignard, Victor: Chemistry, 1912. *P. 192*
Guillaume, Charles Edouard: Physics, 1920. *P. 190*
Gullstrand, Allvar: Medicine, 1911. *P. 193*

H

Haber, Fritz: Chemistry, 1918. *P. 191*
Hahn, Otto: Chemistry, 1944. *P. 190*
Harden, Arthur: Chemistry, 1929. *P. 192*
Haworth, Walter Norman: Chemistry, 1937. *P. 192*
Heisenberg, Werner: Physics, 1932. *P. 189*
Hench, Philip Showalter: Medicine, 1950. *P. 192*
Hertz, Gustav: Physics, 1925. *P. 189*
Hess, Victor Franz: Physics, 1936. *P. 190*
Hess, Walter Rudolf: Medicine, 1949. *P. 193*
Hevesy, George de: Chemistry, 1943. *P. 192*
Heymans, Corneille Jean François: Medicine, 1938. *P. 193*
Heyrovský, Jaroslav: Chemistry, 1959. *P. 191*
Hill, Archibald Vivian: Medicine, 1922. *P. 193*
Hinshelwood, Sir Cyril Norman: Chemistry, 1956. *P. 191*
Hodgkin, Alan Lloyd: Medicine, 1963. *P. 193*
Hoff, Jacobus Henricus van't: Chemistry, 1901. *P. 191*
Hofstadter, Robert: Physics, 1961. *P. 190*
Hopkins, Sir Frederick Gowland: Medicine, 1929. *P. 192*
Houssay, Bernardo Alberto: Medicine, 1947. *P. 192*
Huxley, Andrew Fielding: Medicine, 1963. *P. 193*

J

Jensen, J. Hans D.: Physics, 1963. *P. 190*
Joliot, Frédéric: Chemistry, 1935. *P. 191*
Joliot-Curie, Irène: Chemistry, 1935. *P. 191*

K

Kamerlingh-Onnes, Heike: Physics, 1913. *P. 191*
Karrer, Paul: Chemistry, 1937. *P. 192*

Kendall, Edward Calvin: Medicine, 1950. *P. 192*
Kendrew, John Cowdery: Chemistry, 1962. *P. 193*
Koch, Robert: Medicine, 1905. *P. 194*
Kocher, Emil Theodor: Medicine, 1909. *P. 194*
Kornberg, Arthur: Medicine, 1959. *P. 193*
Kossel, Albrecht: Medicine, 1910. *P. 192*
Krebs, Hans Adolf: Medicine, 1953. *P. 193*
Krogh, Schack August Steenberger: Medicine, 1920. *P. 193*
Kuhn, Richard: Chemistry, 1938. *P. 192*
Kusch, Polykarp: Physics, 1955. *P. 190*

L

Lamb, Willis Eugene: Physics, 1955. *P. 190*
Landau, Lev Davidovič: Physics, 1962. *P. 191*
Landsteiner, Karl: Medicine, 1930. *P. 193*
Langmuir, Irving: Chemistry, 1932. *P. 191*
Laue, Max von: Physics, 1914. *P. 190*
Laveran, Charles Louis Alphonse: Medicine, 1907. *P. 194*
Lawrence, Ernest Orlando: Physics, 1939. *P. 190*
Lederberg, Joshua: Medicine, 1958. *P. 193*
Lee, Tsung Dao: Physics, 1957. *P. 190*
Lenard, Philipp Eduard Anton von: Physics, 1905. *P. 189*
Libby, Willard Frank: Chemistry, 1960. *P. 191*
Lipmann, Fritz Albert: Medicine, 1953. *P. 193*
Lippmann, Gabriel: Physics, 1908. *P. 191*
Loewi, Otto: Medicine, 1936. *P. 193*
Lorentz, Hendrik Antoon: Physics, 1902. *P. 189*

M

Macleod, John James Richard: Medicine, 1923. *P. 192*
Marconi, Guglielmo: Physics, 1909. *P. 189*
Martin, Archer John Porter: Chemistry, 1952. *P. 192*
Mayer, Maria Goeppert: Physics, 1963. *P. 190*
McMillan, Edwin Mattison: Chemistry, 1951. *P. 191*
Mečnikov, Il'ja Il'jič: Medicine, 1908. *P. 194*
Medawar, Peter Brian: Medicine, 1960. *P. 194*
Meyerhof, Otto Fritz: Medicine, 1922. *P. 193*
Michelson, Albert Abraham: Physics, 1907. *P. 190*
Millikan, Robert Andrews: Physics, 1923. *P. 189*
Minot, George Richards: Medicine, 1934. *P. 194*
Moissan, Henri: Chemistry, 1906. *P. 191*
Moniz, Antonio Caetano de Abreu Freire Egas: Medicine, 1949. *P. 193*
Morgan, Thomas Hunt: Medicine, 1933. *P. 193*
Mössbauer, Rudolf Ludwig: Physics, 1961. *P. 190*
Muller, Hermann Joseph: Medicine, 1946. *P. 193*
Müller, Paul Hermann: Medicine, 1948. *P. 194*
Murphy, William Parry: Medicine, 1934. *P. 194*

N

Natta, Giulio: Chemistry, 1963. *P. 192*
Nernst, Walther Hermann: Chemistry, 1920. *P. 191*
Nicolle, Charles Jules Henri: Medicine, 1928. *P. 194*
Northrop, John Howard: Chemistry, 1946. *P. 192*

O

Ochoa, Severo: Medicine, 1959. *P. 193*
Ostwald, Wilhelm: Chemistry, 1909. *P. 191*

P

Pauli, Wolfgang: Physics, 1945. *P. 189*
Pauling, Linus Carl: Chemistry, 1954. *P. 191*
Pavlov, Ivan Petrovič: Medicine, 1904. *P. 193*
Perrin, Jean Baptiste: Physics, 1926. *P. 191*
Perutz, Max Ferdinand: Chemistry, 1962. *P. 193*
Planck, Max Karl Ernst Ludwig: Physics, 1918. *P. 189*
Powell, Cecil Frank: Physics, 1950. *P. 190*
Pregl, Fritz: Chemistry, 1923. *P. 192*
Purcell, Edward Mills: Physics, 1952. *P. 190*

R

Rabi, Isidor Isaac: Physics, 1944. *P. 190*
Raman, Sir Chandrasekhara Venkata: Physics, 1930. *P. 191*
Ramon y Cajal, Santiago: Medicine, 1906. *P. 193*
Ramsay, Sir William: Chemistry, 1904. *P. 191*
Rayleigh, Lord (John William Strutt): Physics, 1904. *P. 191*
Reichstein, Tadeus: Medicine, 1950. *P. 192*

Richards, Dickinson W.: Medicine, 1956. *P. 194*
Richards, Theodore William: Chemistry, 1914. *P. 191*
Richardson, Owen Willans: Physics, 1928. *P. 190*
Richet, Charles Robert: Medicine, 1913. *P. 194*
Robbins, Frederick Chapman: Medicine, 1954. *P. 194*
Robinson, Sir Robert: Chemistry, 1947. *P. 192*
Röntgen, Wilhelm Conrad: Physics, 1901. *P. 189*
Ross, Sir Ronald: Medicine, 1902. *P. 194*
Rutherford, Ernest: Chemistry, 1908. *P. 191*
Ružička, Leopold: Chemistry, 1939. *P. 192*

S

Sabatier, Paul: Chemistry, 1912. *P. 192*
Sanger, Frederick: Chemistry, 1958. *P. 193*
Schrödinger, Erwin: Physics, 1933. *P. 189*
Seaborg, Glenn Theodore: Chemistry, 1951. *P. 191*
Segrè, Emilio Gino: Physics, 1959. *P. 190*
Semenov, Nikolaj Nikolajevič: Chemistry, 1956. *P. 191*
Sherrington, Sir Charles Scott: Medicine, 1932. *P. 193*
Shockley, William: Physics, 1956. *P. 190*
Siegbahn, Karl Manne Georg: Physics, 1924. *P. 189*
Soddy, Frederick: Chemistry, 1921. *P. 191*
Spemann, Hans: Medicine, 1935. *P. 193*
Stanley, Wendell Meredith: Chemistry, 1946. *P. 192*
Stark, Johannes: Physics, 1919. *P. 189*
Staudinger, Hermann: Chemistry, 1953. *P. 192*
Stern, Otto: Physics, 1943. *P. 190*
Sumner, James Batcheller: Chemistry, 1946. *P. 192*
Svedberg, The: Chemistry, 1926. *P. 191*
Synge, Richard Laurence Millington: Chemistry, 1952. *P. 193*
Szent-Györgyi von Nagyrapolt, Albert: Medicine, 1937. *P. 192*

T

Tamm, Igor' Jevgen'evič: Physics, 1958. *P. 190*
Tatum, Edward Lawrie: Medicine, 1958. *P. 193*
Theiler, Max: Medicine, 1951. *P. 194*
Theorell, Axel Hugo Theodor: Medicine, 1955. *P. 193*
Thomson, George Paget: Physics, 1937. *P. 190*
Thomson, Joseph John: Physics, 1906. *P. 189*
Tiselius, Arne Wilhelm Kaurin: Chemistry, 1948. *P. 192*
Todd, Sir Alexander Robertus: Chemistry, 1957. *P. 193*

U

Urey, Harold Clayton: Chemistry, 1934. *P. 191*

V

Vigneaud, Vincent du: Chemistry, 1955. *P. 193*
Virtanen, Artturi Ilmari: Chemistry, 1945. *P. 192*

W

Waals, Johannes Diderik van der: Physics, 1910. *P. 191*
Wagner-Jauregg, Julius: Medicine, 1927. *P. 194*
Waksman, Selman Abraham: Medicine, 1952. *P. 194*
Wallach, Otto: Chemistry, 1910. *P. 192*
Walton, Ernest Thomas Sinton: Physics, 1951. *P. 190*
Warburg, Otto Heinrich: Medicine, 1931. *P. 192*
Watson, James Dewey: Medicine, 1962. *P. 193*
Weller, Thomas Huckle: Medicine, 1954. *P. 194*
Werner, Alfred: Chemistry, 1913. *P. 191*
Whipple, George Hoyt: Medicine, 1934. *P. 194*
Wieland, Heinrich Otto: Chemistry, 1927. *P. 192*
Wien, Wilhelm: Physics, 1911. *P. 191*
Wigner, Eugene Paul: Physics, 1963. *P. 190*
Wilkins, Maurice Hugh Frederick: Medicine, 1962. *P. 193*
Willstätter, Richard Martin: Chemistry, 1915. *P. 192*
Wilson, Charles Thomson Rees: Physics, 1927. *P. 190*
Windaus, Adolf Otto Reinhold: Chemistry, 1928. *P. 192*

Y

Yang, Chen Ning: Physics, 1957. *P. 190*
Yukawa, Hideki: Physics, 1949. *P. 190*

Z

Zeeman, Pieter: Physics, 1902. *P. 189*
Zernike, Frits: Physics, 1953. *P. 190*
Ziegler, Karl: Chemistry, 1963. *P. 192*
Zsigmondy, Richard Adolf: Chemistry, 1925. *P. 191*

BIBLIOGRAPHY

Scientific Personality

*Roe, Anne, *The Making of a Scientist.* Dodd, Mead, 1952.
Taylor, Calvin W., and Frank Barron, eds., *Scientific Creativity.* John Wiley & Sons, 1963.

Biography

Bergengren, Erik, *Alfred Nobel.* Thomas Nelson & Sons, 1962.
*Editors of FORTUNE, *Great American Scientists.* Prentice-Hall, 1961.
Schück, H., R. Sohlman, A. Österling, G. Liljestrand, A. Westgren, M. Siegbahn, A. Schou and N. K. Stahle, *Nobel, The Man and His Prizes,* ed. by the Nobel Foundation. Elsevier, 1962.

General History of Science

†Butterfield, Herbert, *The Origins of Modern Science.* Macmillan, 1961.
Dampier, Sir William Cecil, *A History of Science.* Cambridge University Press, 1961.
†Mason, Stephen G., *A History of the Sciences.* Collier Books, 1962.
*Price, Derek J. de Solla, *Science Since Babylon.* Yale University Press, 1961.
Sarton, George, *A History of Science* (2 vols.) Harvard University Press, 1959.
Taton, René, *Ancient and Medieval Science.* Basic Books, 1963.
*Wolf, A., *A History of Science, Technology and Philosophy in the 16th and 17th Centuries* (2 vols.). Peter Smith, 1963. *A History of Science, Technology and Philosophy in the 18th Century* (2 vols.). Peter Smith, 1963.

Histories of Special Fields

*Adams, Frank Dawson, *The Birth and Development of the Geological Sciences.* Peter Smith, 1938.
Hoyle, Fred, *Astronomy.* Doubleday, 1962.
Gamow, George, *Biography of Physics.* Harper & Brothers, 1961.
Leicester, Henry M., *The Historical Background of Chemistry.* John Wiley & Sons, 1956.
Schneer, Cecil, *The Search for Order,* Harper & Brothers, 1960.
†Struik, Dirk, J. A., *A Concise History of Mathematics.* Dover, 1948.
Taylor, Gordon Rattray, *The Science of Life.* McGraw-Hill, 1963.

Philosophy and Methodology

†Bronowski, J., *Science and Human Values.* Harper & Brothers, 1959.
†Margenau, Henry, *The Nature of Physical Reality.* McGraw-Hill, 1959.
*Margenau, Henry, *Open Vistas.* Yale University Press, 1961.
†Northrop, F.S.C., *The Logic of the Sciences and the Humanities.* Meridian, 1959.
*Platt, John Rader, *The Excitement of Science.* Houghton Mifflin, 1962.
†Poincaré, Henri, *Science and Method.* Dover Publications, 1914.
†Ritchie, A. D., *Scientific Method.* Littlefield, 1960.
†Sullivan, J.W.N., *The Limitations of Science.* New American Library, 1949.

Science and Modern Life

Barber, Bernard, and Walter Hirsch, *The Sociology of Science.* Free Press of Glencoe, 1962.
Barzun, Jacques, *Science: The Glorious Entertainment.* Harper & Row, 1964.
Dupree, A. Hunter, *Science in the Federal Government.* Harvard University Press, 1957.
Kepes, Gyorgy, *The New Landscape in Art and Science.* Paul Theobald, 1956.
Lindsay, Robert B., *The Role of Science in Civilization.* Harper & Row, 1963.
†Obler, Paul, and Herman Estrin, eds., *The New Scientist,* Doubleday, 1962.
Piel, Gerard, *Science in the Cause of Man.* Alfred A. Knopf, 1962.
*Snow, C. P., *The Two Cultures: and A Second Look.* Cambridge University Press, 1964.
Wolfle, Dael, *Science and Public Policy.* University of Nebraska, 1959.

---

*Available also in paperback edition.
†Available only in paperback edition.

ACKNOWLEDGMENTS

The editors of this book are indebted to the following people and institutions: Mrs. Kathryn S. Arnow, Thomas J. Mills, Office of Economic and Manpower Studies, and Dr. Henry Birnbaum, Department Head, Office of Science Information Services, National Science Foundation, Washington, D.C.; Sylvio Bedini, Curator, Division of Mechanical and Civil Engineering, Smithsonian Institution, Washington, D.C.; Robert W. Beyers, Director, and Robert Lamar, Stanford University News Service, Palo Alto, Calif.; L. Sprague de Camp; Dr. Willard D. Cheek, Senior Physicist, General Motors Research Laboratories, Warren, Mich.; Sylvan Cole Jr., President, Associated American Artists, Inc., New York City; Ray Colvig and Daniel Wilkes, Public Information Office, University of California, Berkeley; Harold Dorn, Lecturer in the History of Mathematics, The City College of New York; Bern Dibner, Curator, The Burndy Library, Norwalk, Conn.; Dr. William B. Fowler, Paul A. Pion, Dr. Nicholas Samios, Dr. Ralph P. Shutt, Brookhaven National Laboratories, Upton, N.Y.; Dr. Renée Fox, Associate Professor of Sociology, Barnard College, Columbia University; Dr. Murray Gell-Mann, Professor of Physics, Dr. Yuval Ne'eman, Visiting Professor of Theoretical Physics, James R. Miller, Director, Public Relations, California Institute of Technology; Hugo Gernsback; Dr. Daniel S. Greenberg, Associate Professor of History, Columbia University; Dr. Fred L. Holmes, Assistant Professor of History of Science, Massachusetts Institute of Technology; Romana Javitz, Picture Collection, Lewis M. Stark, Director, and Maud D. Cole, Rare Book Division, New York Public Library; Dr. Gyorgy Kepes, Professor of Architecture, Massachusetts Institute of Technology; Frederick G. Kilgour, Librarian of the Medical Library, Yale University; Eugene H. Kone, Director, and Audrey Meyers, Assistant to Director, Office of Public Relations, American Institute of Physics, New York City; Dr. Melvin Kranzberg, Professor of History, Case Institute of Technology; Samuel Moskowitz; Dr. Glenn Negley, Professor of Philosophy, Duke University; the Nobel Foundation, Stockholm, Sweden; Dr. Derek J. de Solla Price, Chairman, and Elizabeth H. Thomson, Research Assistant, Department of History of Science and Medicine, Yale University; Geoffrey Smithers, Professor of English Language, University of Durham, Durham, England; Paul West, Office of Publications and Public Information, University of California, San Diego; Dr. L. Pearce Williams, Associate Professor of History of Science, Cornell University; Dr. Harold Woolf, Professor of History of Science, The Johns Hopkins University.

INDEX

Numerals in italics indicate a photograph or painting of the subject mentioned.

A

"Absolute zero," 56
Abstracts, journals of, 108, *109*
Académie des Sciences, *74*, 126
Aerospace research and industry, California, 130
Agricola, Georgius, *94*
AGS (alternating gradient synchrotron), *diagram* 66
Air Force Office of Scientific Research, 127
Air pumps, *44-45*
Alchemy, 46, 47, 79, 82, *90*, 91
Alder, Kurt, *175*
Alexander the Great, 79, 82
Alexandria, Egypt, "Museum" at, 79-80
Algebra, 78, 80, 86, 100
Aiken, Henry, 155
Alpher, Ralph, 13-14
Alternating gradient synchrotron, *diagram* 66
Alvarez, Luis, 14
American Academy of Arts and Sciences, 125
American Association for the Advancement of Science, 31, 125-126
American Chemical Society, 126
American Institute of Physics, 130
American Journal of Science, 106
American Miscellaneous Society (AMSOC), 12-13
American Philosophical Society, 125
American Physical Society, 72
Ammonia clock, *55*
Analysis: chemical, 46; logical, 56-57, 77; mathematical discipline, 86, 100; numerical, 78; statistical, 84
Analytic chemistry, 91, 100
Analytic geometry, 86, 87, 100
Anatomy, 76, 96, 101; comparative, 97, 101; historic roots of, 78
Animals, classification of, 82, 97
Anthropologists, I.Q.—test results, *34*
Anthropology, 98, 99, 101
Anticoagulant drugs, research, *50*
Applied science, 145-146; expenditures, *graph* 124. *See also* Technology
Aptitude tests of scientists, 32; problems and answers, *32*, *34*
Arab contributions to science, 80, 92
Archeology, 76, 77, 98, 99, 101, *114*
Archimedes, 52, 151
Architecture, impact of science on, 166, 167
Argonne National Laboratories, Chicago, 11, 129
Aristarchus, 80
Aristotle, 53, 59, 79, 96, *104*
Arithmetic, 78, 86, 100, 104
Armillaries, *42-43*; equatorial, 38, *39*
Art, influence of science on, *164*, 165, 166-169
Associated Universities, 128-129
Associations, scientific, 125-126
Astrolabe, *36*, *37*
Astrology, 79
Astronautics, 93, 101
Astronomer(s), *116-117*; dating of term, 29; number of living, 117
Astronomy, 76, 82, 83, 104, 117; *vs.* astrology, 79; beginnings of, 78-79, 92; history of, *92-93*; instruments, *36*, *37-40*, *42-43* (*see also* Telescopes); interdisciplinary specialties, *chart* 100-101; specialties, *chart* 92-93, 101
Astrophysics, 92, 93, 100, 101
Atlantis, research vessel, 128
Atom(s), 48, 83; arrangement in crystals, *24-25*; arrangement on a needle's point, *17*; forces within, 64-65, 72; models of, 61-62, *64*
Atom bomb, development of, 135, 148, *graph* 149
Atom smashers, 72; beginnings of, 48, *49*, *132-133*; Berkeley bevatron, *134*, *135*, 141; Berkeley cyclotrons, *132-135*; Berkeley synchrocyclotron, *135*; Brookhaven synchrotron (AGS), *diagram* 66, 128; Stanford linear accelerator, *140-141*
Atomic Energy Commission, 127, 129, 135
Atomic nucleus, discovery of, 48, 53
Atomic particles. *See* Particles, subatomic
Atomic physics, 88, 89, 100; early instruments, *48-49*
Atomic research, technological applications, *144*, 145, 148
Atomic weight, theory of, 83
Automated science libraries, 108-109
Automation: 1884 cartoon, *151*; examples of future, 148-150
Automobiles, automatic pilots, 149
Aviation, 146; beginnings of, *158*; early cartoons, *150*, *158-159*; predictions for future developments, 149
Azimuth quadrant, 38, *39*

B

Baade, Walter, 10
Babylonians: astrology, 79; contributions to science, 78
Bacon, Francis, 52, 59, 81, 82, 152
Ballistics, 77
Barrow, Isaac, 34
Bayeux Tapestry, *76*
Beadle, George, 14
Bean-root nodule, cross section, *20*
Bell, Alexander Graham, 146, 148
Bell Telephone, laboratories, 129
Benzene ring, *91*
Berkeley, Calif., Lawrence Radiation Laboratory, *132-135*
Bethe, Hans, 14
Bethesda, Md., National Institutes of Health, 127
Bevatron, Berkeley, *134*, *135*, 141
Bioastronautics, 75
Biochemistry, 83, 91, 97, 100, 101, 150
Biological sciences. *See* Life sciences
Biologist(s): dating of term, 29; divorce rate, *33*; I.Q.—test results, *34*
Biology, 76, 78; molecular, 96, 97, 100, 101
Biophysics, 97, 100, 101, 165
Biotron, 128
Black, Joseph, 146
Blumenbach, Johann, 98
Boas, Franz, *99*
Boethius, *104*
Bohr, Niels, model of atom, 61-62
Bondi, Hermann, *117*
Book publishing, science, 106
Bose, Satyendranath, 110, *111*
Botanist, dating of term, 29
Botany, 77, 96, 101
Bowen, Edward, *116*
Boyle, Robert, 44, 81, *91*, 125, 146
Bragg, Sir William Henry, 185
Bragg, William Lawrence, 184, 185
Brahe, Tycho, *38*, 39
Branches of science, 75, 77, 84, *85-101*. *See also* Specialties
Bridgman, Percy, 55
Brion, Marcel, 166
Bronk, Detlev, 120, *121*
Brookhaven National Laboratory, 108, 127, 128; search for Omega Minus at, *66-73*
Bubble chamber, 66, *67-71*, *73*; photographs of particle tracks, *63*, 66, 68, *70-71*
Bundy, McGeorge, 31
Bushnell, David, 160
Byrd, Harry, 127

C

Calcium sulfate crystals, *26-27*
Calculus, 86, 87, 100
California, "the science state," 130, *map* 131, *132-143*
California Institute of Technology, 129, 130, *139*, *142-143*
Camp Irwin, Calif., *map* 131
Carnegie Institution, 129
Carver, George Washington, 14
Cause-and-effect relationship, 61, 77, *172-173*
Cavendish Laboratory, Cambridge, 36, *48-49*, 186
"C-domain," 57. *See also* Concepts
Celestial mechanics, 92, 101
Center for Advanced Study in the Behavioral Sciences, Stanford, 127
Cézanne, Paul, 167
Chaldeans: engineering, 78; necromancy, 79
Challenger, research vessel, *95*
Character traits, scientists', 32-33; anecdotes, 10-16
Chemist, dating of term, 29
Chemistry, 76; defined, 90; early 19th Century Laboratory and tools, *46-47*; historic roots of, 78, 82; history of, 46, *90-91*; interdisciplinary specialties, *chart* 100-101; specialties, 84, *chart* 90-91, 101
Cherwell, Lord, 13
Chiton shell, *18*
Christofilos, Nicholas, 108
Clocks, *42*, *54-55*
Cockcroft, Sir John, 48, 186
Cognitive experience, 54
Color, scientific analysis of, 166
Columbia University, physics library, *102*
Comets, Halley's work on, *76*
Communication, scientific. *See* Information; Language
Communications satellites, 149
Comparative anatomy, 97, 101
"Complex systems," 83-84
Computer libraries: general, 149; science, 108-109
Computers, 84, 86, *87*, 147, 148; predictions for future uses of, 148, 149-150, 151, 170; use in arts and literature, 165, 173
Concepts, scientific, 55-57, 59-62; deduction of theory from, 56-57, *diagram* 58; "operationally defined," 55
Condillac, Etienne de, 105
Congress, U.S., appropriations for science purposes, 125, 126, 127
Constitution, U.S., on promotion of science, 126
Construction materials, future, 149
Copernicus, Nicholas, 80, 92, 93, 170, 171; his diagram of solar system, *92*
Copper crystal, light patterns of groove in, *22*
Copper sulfate crystals, *26-27*
Cortelyou, John, 15
Cosmology, 92, 94, 101
Creativity of scientists, 30, 32
Crick, Francis H. C., 184, *186*
Cryogenics, 75, 89, 100
Cryology, 75
Crystallography, 89
Crystals, *25*; various patterns, *22*, *24-27*
Cultural anthropology, 98, 99, 101
Culture, impact of science on, 145, 165-174
Curie, Marie Sklodowska, 180, 184, 186
Curie, Pierre, 186
Cyclotrons, Berkeley, *132-135*, 141
Cytology, 96, 97, 101, 128

D

Daguerre, Louis Jacques Mandé, 148
Dalton, John, 82
Dante, Alighieri, 166
Dart, Harry Grant, 158, 159
Darwin, Charles, theory of evolution, 53, 64, 83, 97, 106, 170; mentioned, 30, 155
Darwin, Erasmus, 155
Data, scientific, defined, 54
Da Vinci, Leonardo, 104
De Broglie, Louis-Victor, 184
Deductive logic, 56-57, 79
De Forest, Lee, 136
Degrees, Ph.D., statistics, 30
DeMaeyer, Leo C. M., 112
Descartes, René, 59, 81, 82, 86
Descriptive sciences, 76
Design, influence of science on, 167
Development, scientific: financing, *graphs* 124-125; period between discovery and its application, 147-148, *graph* 149
Dielectrics, 75
Diels, Otto Paul, *175*
Differential equations, 87, 100
Dilworth, Robert P., *139*
Discovery, scientific, 52-53; duplications of, 108, 152
Divorce rates of scientists, 33
DNA molecule, 57, 96
Dynamite, invention of, 177

E

Earle, Wilton, 53
Earth sciences: history of, *94-95*; interdisciplinary specialties, *chart* 100-101; specialties, *chart* 94-95, 101
Eaton, Amos, 106
Eccles, Sir John Carew, *182*
Ecology, 97, 101
Economics, 98, 99, 101, 169
Eddington, Arthur, 106
Edison, Thomas Alva, 146
Education, scientific: Alexandrian "Museum," 79-80; fund sources and expenditures, *graph* 125; medieval, 104; modern, 103, *138-139*, 142; need for cross-fertilization of sciences and humanities, 109-110. *See also* individual universities
Edwards, Calif., *map* 131
Eel scales, *20*
Egypt, ancient: alchemy, 79; contributions to science, 78, *90*
Eigen, Manfred, *112*
Eightfold way, in particle physics, 64, *65*, 72
Einstein, Albert, 9, 53, 57, 58, 83, *89*, 110, *111*; Nobel Prize medal, *185*; theory of relativity, 89, 165; work on unified field theory, 72, 83; quoted, 77, 173; mentioned, 13, 30, 51, 87
Electric horse, *156*
Electric light, 146; of future, 149
Electric motor, 151, 152
Electricity, 45, 89, 100, 148, 156; unfulfilled prophesies for use of, *156-157*
Electromagnetic force, 64, 65, 72
Electromagnetic spectrum, 83, 147
Electromagnetism, 64, 72, *83*, 88, 146; discovery, 148; Maxwell's theory, 53, 57, 83
Electron, *64*, 146; concepts of, 59-62; discovery of, 48; location of, probability chart, *61*
Electron microscope, 136
Electronic music, 168; score, *168*
Electronics, applied, 146
Electronics industry, California, 130, 136, *137*
Elements, 145; periodic table of, 83, 91
Elements, Euclid, 80, 86
Embryology, 101; beginnings of, 79, 96
Empiricism *vs.* rationalism, 59
Engineering, historic roots of, 78
Engineers, number of, in U.S., *graph* 31, 124
Enlightenment, age of, 81-82
Epistemology, 52
Eratosthenes, 80, *94*
Euclid, 80, *86*, *104*
Euclidean geometry, 77, 86
Evolution, Darwin's theory of, 53, 64, 83, 97, 106, 170
Exact sciences, 77
Expenditures: for applied research, 124; for basic research, 124, 127, 131; for development, 124; total U.S. scientific activity, *graphs* 124-125
Experimentation: early history, 79-80; pragmatic, 77; systematic, 77
Exploration, 94, 98

F

Fact, scientific, 53-55, 58-59; interaction with theory, 53-58, *diagram* 58; protocol facts, 54-55, *diagram* 58; qualitative, 55; quantitative, 55
Fahrenheit scale of temperature, 56
Faraday, Michael, 152
Fermat, Pierre de, 86
Feynman, Richard P., 11-12
Field ion microscope, *16*
Financing. *See* Expenditures; Funds
Fleming, Alexander, 52
Fleming, J. A., 146
Force: electromagnetic, 64, 65, 72; strong (atomic nucleus), 64, 65, 72
Ford Foundation, 129
"Form follows function," concept, 167
Formulas, expression of scientific laws by, 55
Franklin, Benjamin, 120, 125, 126
Freud, Sigmund, 83, 170
Functionalism, 167
Functions, theory of, 87, 100
Fundamental numbers, 56
Funds for R & D, sources of, *graph* 125, 128-129

G

Galileo Galilei, 40, 52-53, 54, 56, 81, *88*, *93*, 128; and free fall, 58, 86, 88
Gamow, George, 13-14
Gases, basic law of, 55-56
Gauss, Karl Friedrich, 86
Gell-Mann, Murray, *64*, 65, *72*
Genetics, 97, 100, 136, 147, 150
Geochemistry, 95, 100, 101
Geodesy, 94, 100, 101
Geography, 82; physical, 95, 101
Geologist, dating of term, 29
Geology, 76, 82, 95, 101

197

Geometry, 79, 86, 100, 104; Euclidean, 77, 86; non-Euclidean, 57, 86, 87, 100
Geomorphology, 95, 101
Geophysics, 95, 100, 101, 120
Gernsback, Hugo, 162
Gibbs, Josiah Willard, 146-147
Gilbert, William, 56, 81
Glass, light-sensitive, 149, 150
Goddard, Robert, 15
Government, influence of scientist on, 169-170
Government, study of, 98
Government, U.S., role in scientific endeavor, 123, *graph* 125, 126-127, 128-129
Granger, John V. N., *137*
Gravitation, 56, 58; Newton's law of, 52, 58, 81, 89
Greek science, 77, 79-80, 86, 92, 96
Green Bank, W. Va., observatory, 128-129
Greenough, Horatio, 167
Grimm brothers, 83
Groves, General Leslie R., 30
Guericke, Otto von, 44, 45
Guppy, Nicholas, 15-16
Gutenberg, Beno, 13

H

Haldane, J.B.S., 58
Halley, Edmund, 76, 81
Halley's comet, 76
Harvard University, 109
Harvey, William, 81, 96
Hawkins, Gerald S., 78
Haydn, Joseph, 166
Heat, 88, 148. *See also* Thermodynamics
Heisenberg, Werner, 61, 172-173
Helicopters, future use of, 149
Helmont, Johann Baptista van, 90
Henry, Joseph, 151, 152
Heredity, 55
Hero of Alexandria, *80, 88*, 147
Herophilus, 80
Herschel, William, 93
Hertz, Heinrich Rudolph, 57
Hewlett, William R., *137*
Hexagon, as basic natural shape, *20-21*
High-energy physics, 62, 72, 141. *See also* Atom smashers
"High-technology" industries, *136-137*
Hindemith, Paul, 168
Hippocrates, *96*
Histology, 96, 97, 101
Historical geology, 95, 101
History: classification in body of knowledge, 76, 98; use of scientific evidence in, 165
Hittites, contributions to science, 78
Hodgkin, Alan, 182, *186*
Honeycomb, *21*
Hooke, Robert, 81, 125, 146; his *Micrographia, 40;* his microscope, *40, 97*
Hoover, Herbert, 136
Huling, Richard, *28*
Humanists: defined, 103; *vs.* scientists, 103-104, 105, 106, 117, 165
Humanities: defined, 103; need for cross-fertilization with sciences, 109-110, 173
Hume, David, 59
Hutton, James, 82, 95
Huxley, Andrew, 182
Huxley, Julian, 106, 172
Huxley, Thomas, 106, 109
Huygens, Christian, 81, 146
Hydrology, 95, 100, 101
Hydroplanes, future use of, 149
Hypothesis, forming of, as step in scientific method, 52, 57

I

Idea, scientific: interaction with fact, 51, 53, 145; rationalist-empiricist dispute, 59; *vs.* reality, in advanced science, 58-62
IGY (International Geophysical Year), 120
Impetus School, 53
Indians, ancient, contributions to science, 78
Inductive logic, 56
Industrial Revolution, 148, 155
Industry, role in scientific endeavor, 123-124, *graph* 125, 127, 129, 136, *137*
Information, scientific, dissemination of, 103-110; automated (computer) library, 108-109; early records, 78, 81; impact of invention of printing, 81; international cooperation, 35, 82, 108, 110, 120; 19th Century popular lecturing, 106; popular science reporting, 106-107; scientific papers and abstracts, 107-108, *graph* 109; word-of-mouth (conferences, etc.), 108

Information theory, 86, 87, 100, 147
Inorganic chemistry, 90, 91, 100
Institute for Advanced Study, Princeton, 10, 127
Institutes and laboratories, 125, 127-129
Instruments, 36, *37-49;* astronomical, *37-40, 42-43* (see also Telescopes); of early atomic physics, *48-49;* 19th Century chemistry lab, *46-47;* of Renaissance, 36, *42-43. See also* Atom smashers; Microscopes
Intelligence tests of scientists, 32; problems and results, *32, 34*
Interdisciplinary sciences, 75, *chart* 100-101
International Business Machines, 129
International Geophysical Year, 120
International scientific discourse and cooperation, 35, 82, 108, 110, 112-113, 120
International Year of the Quiet Sun, 1964-1965, 120
Intuitionism, *75*
Inventions, 146-148; duplications of, 152; predictions for future, 148-150; predictions of past fulfilled, 148, *150-151*, 152, *153;* sociological impact of, 145, 150-151, 155
Iron crystals, *24-25*
Ismail, Abdel Aziz, *118*

J

James, William, 15
Jeans, James, 106
Jefferson, Thomas, 98, 126, 151
Jensen, J. Hans D., 179, *180-181*
Jet Propulsion Laboratory, 129, 139
Journalism, science, 106-107
Journals, scientific, *102*, 107-108, *graph* 109; of abstracts, 108, 109
Junto (*later* American Philosophical Society), 125

K

Kaisel, Stanley F., *137*
Kandinsky, Vasili, 166
Kant, Immanuel, 59
Kapitza, Peter, *112*
Kaplan, Joseph, *120*
Kappa Minus particle, *68*
Kekulé, Friedrich, 91
Kelvin scale of temperature, 56
Kepes, Gyorgy, 16
Kepler, Johannes, 38, 43, 56; laws of planetary motion, 53, 81
Klystron tube, 136
Koch, Robert, 83

L

Laboratories and institutes, 36, *46-47*, 125, 127-129. *See also individual laboratories, e.g.,* Brookhaven National Laboratory; Cavendish Laboratory; Lawrence Radiation Laboratory; Stanford Research Institute
Lamarck, Jean-Baptiste, 167
Language, scientific, 35, 104-106, *examples* 107
Langue des Calculs, Condillac, 105
Laplace, Pierre Simon de, 172
Laser beam, *136-137*, 149, *153*
Lavoisier, Antoine, *35*, 82, 90; apparatus of, *91*
Lawrence, Ernest O., *132, 133*
Lawrence Radiation Laboratory, 129, *132-135*
Lederberg, Joshua, *136*
Lee, Tsung Dao, 8, 186
Leeuwenhoek, Anton van, 81
Leibnitz, Gottfried Wilhelm von, 86, *87*
Leloir, Luis Federico, *113*
"Leviathan of Parsontown," telescope, 128
Lewis and Clark Expedition, 126
Libby, Willard, *138*
Liberal arts, 104
Libraries: computer, 149; "phonographic," *156;* science, *102;* science, automated, 108-109
Liebig, Justus von, 46, 47
Life on other planets, probability of, 171, *chart* 172
Life (biological) sciences, 75, 77, 83; history of, *96-97;* interdisciplinary specialties, *chart* 100-101; specialties, *chart* 96-97, 101
Life scientists, number in U.S., 124
Light: bending of (theory of Relativity), *diagram* 57; eletromagnetic nature of, 83, 88; patterns, *22-23;* speed of, 56
Light quanta, 64
Linear accelerator, Stanford, *140-141*
Linnaeus (Carl von Linné), *82, 97*
Literature, use of scientific aids in, 168
Livermore, Calif., 135

Locke, John, 99
Logic, 86, 100; deductive, 56-57, 79
Logic, mathematical, 87, 100
Logical analysis, 56-57, 77; and proof, 57-58, 77
Lopez-Ibor, Juan, *113*
Lorenz, Konrad, 13
Los Alamos, 30, 127
Los Angeles, as center of science, 130, *map* 131, *138-139*
Lunar and Planetary Laboratory, University of Arizona, 129
Lunts, A. G., 108
Luther, Martin, 171

M

Machiavelli, Niccolò, *98*
Machine tooling, beginnings of, 42, 43
McLellan, William H., 12
Madison, James, 126
Magdeburg Hemispheres, 45
Magnetism, 89, 100. *See also* Electromagnetism
Majaj, Amin, *118*
Manhattan Project, 135
Marais, Marin, 166
Marconi, Guglielmo, 57
Marlowe, Christopher, 166
Marx, Karl, 98, *99*
Massachusetts Institute of Technology, 109, *127*
Mathematical aptitude of scientists, 32; test problem and answer, *32, 34*
Mathematical logic, 87, 100
Mathematician(s): dating of term, 29; number in U.S., 124
Mathematics, 75, 77, 147; branches and specialties, *chart* 86-87, 100; history of, *86-87;* as language of science, 104-105
Maxwell, James Clerk, 53, 57, 83, 88
Mayer, Maria Goeppert, *140*, 179, *180-181*
Mead, Margaret, 31
Measurement, 81; as step in scientific method, 52, 54-56, *diagram* 58; of time, advances in, *54-55;* units of, 56
Mechanics, 80, 88, 100, 148; statistical, 89, 100
Medawar, Peter, 186
Medical chemistry, 90, 91
Medical physics, 132
Medical science, 100, 101, 115, 118; early, 78, 80, *96;* Harvey, 96
Medical scientists, *114, 118-119*
Mendel, Gregor, 55
Mendeleyev, Dmitri, 82, 91
Mentality, scientists', types of, 30, 51
Mercury vapor-tube light, 136
Mesomereology, 75
Mesons, 64, 65
Mesopotamia, contributions to science, 78
Metallurgy, 146, 149; historic roots of, 78
Meteorology, 76, 94, 100, 101
Method. *See* Scientific method
Métraux, Rhoda, 31
Metric system, 56
Microbiology, 96, 97, 101
Micrographia, Hooke, 40
Microscopes, 36, *41, 96, 153;* electron, 136; field ion, 16; Hooke's, *40, 97;* invention of, 40
Microwave Electronics Corporation, 137
Middle Ages: education, 104; science and technology of, 80-81, 90, 92, 96
Milton, John, 166
Minakami, Takeshi, *115*
Mineralogy, 82, 88, 94, 95, 101
Miro, Joan, 167
Mofidi, Chamseddine, *119*
Mohole, project, 12-13
Molecular biology, 96, 97, 100, 101
Molecular Medicine, Lieutenant Joseph P. Kennedy Jr. Laboratory for, 136
Molecular physics, 88, 89, 100
Molecules, 83, 145; motion of, 56, 88; understanding of, and technological applications, 145, 148, 149
Mondrian, Piet, 167
Monge, Carlos, *114, 115*
More, Thomas, *98*
Morgan, Thomas Hunt, 11
Morse, Samuel F. B., 146
Mosquito eye, *20-21*
Motion, Newton's laws of, 52, 89, 145
Motion pictures, 146
Mount Palomar Observatory, 127, 129, *map* 131; 200-inch telescope, 128, 133
Mount Wilson Observatory, 127, 129; 100-inch telescope, 128
Müller, Erwin, 16
Multivator, *136*
Munk, Walter, 12
Mushroom cap, *18*
Music, impact of science on, 166, 167-168, 169

N

National Academy of Science, 120, 126
National Aeronautics and Space Administration (NASA), 109, 129
National Association of Science Writers, 106
National Institutes of Health, Bethesda, Md., 127
National Institutes of Mental Health, 127
National Library of Medicine, 109
National Radio Astronomy Observatory, 128-129
National Research Council, 126
National Science Foundation, 13, 32, 75, 98, 127, 129
Nature, order in. *See* Order
Necromancy, 79
Ne'eman, Yuval, *64*, 65
Neurology, 83
Neurophysiology, 147
Neutron, *64*
New Atlantis, Bacon, 152
Newcomen, Thomas, 146
Newspapers, science reporting in, 106-107
Newton, Isaac, 34, *87, 89*, 125; invention of calculus, 86, 87; laws of motion and gravitation, 52, 56, 58, 81, 89, 145; his telescope, *40;* mentioned, 51, 82, 95
Nobel, Alfred, 174, *176*, 177, 180, 186; his will, 177, *178*
Nobel, Immanuel, *177*
Nobel Foundation, 179, 180, 183
Nobel Prize, 174, 177; award ceremony, banquet and ball, 180-183, *186-187;* diplomas, *184*, 185; medal, *185;* money, 174, 186
Nobel Prize winners, *48, 133, 138, 140, 174, 175, 180-183, 186-187;* in science, California's share, 130; selection of, 174, 178, *179*
Non-Euclidean geometry, 57, 86, 87, 100
Nuclear chemistry, 91, 100
Nuclear particles. *See* Particles, subatomic
Nuclear physics, 88, 89, 100
Nuclear-powered shipping, 144
Number systems, 78
Number theory, 86, 100
Numbers, fundamental, 56
Numerical analysis, 78
Numerical *vs.* verbal definition, 55

O

Oak Ridge, 127
Observation, as step in scientific method, 52, 54
Observational scientists, 30, 51, 82
Ocean mining, 150, 160
Oceanography, 76, 94, 95, 100, 101, 128, 141
Oersted, Hans Christian, 148
Office of Naval Research, 127
Office of Scientific Research and Development, 126
Ölander, Arne, *179*
Omega Minus particle: prediction of existence of, 62, 64-65; search and discovery, *66-73;* track, *63, 70-71*
Operationally defined concepts, 55
Optics, 79, 88, 100
Order in nature, 16, *17-21, 24-25;* disruption, *26-27*
Oresme, Nicole, 88
Organ, Hero's wind-powered, *80*
Organic chemistry, 46, 91, 100
Organizations, scientific, 125-129
Origin of Chemical Elements, The, Alpher, Bethe and Gamow, 14
Otis brothers, 151

P

Packard, David, *137*
Painting, impact of science on, *164*, 166-167, 168, 169
Paleontology, 76, 82, 95, 101
Palo Alto, Calif., *136-137*
Palomar Observatory. *See* Mount Palomar
Papin, Denis, 146
Paracelsus, *90*
Parity law, investigation of, *8*, 186
Particle accelerators. *See* Atom smashers
Particle physics, 88, 89, 100
Particles, subatomic, 59, 61, 62, 72, 83, 134; defined, 64; enumeration, 64, 65; multiplets of, 65; properties of, and their rating, 64, 65; research, 48, 62, *63-73,* 133 (see also Atom smashers); short-lived, 64; supermultiplets of, 65. *See also* Electron; Omega Minus
Pascal, Blaise, 86
Pasteur, Louis, 83, 109
Pathology, 76, 96, 101
Paul, Frank, 162

Pavlov, Ivan Petrovič, 184
Peabody, Francis Weld, 15
Pedal planes, 149, 150
Pendulum clock, 54
Pepys, Samuel, 125
Periodic table of elements, 83, 91
Periodicals, scientific. See Journals
Personality, scientists', 32-33; anecdotes, 10-16
Petrology, 95, 101
Pharmacology, 91, 100, 101
Philosophy, 98
Phoenicians, contributions to science, 78
"Phonographic library," 156
Photography, period of development, 148, graph 149
Photon, 64, 65
Physical anthropology, 98, 99, 101
Physical astronomy, 92, 93, 100, 101
Physical chemistry, 91, 100
Physical geography, 95, 101
Physical Review, The, 14, 129
Physical sciences, 75, 77, 83
Physical scientists, number in U.S., 124
Physicist(s): dating and origin of term, 29; divorce rate, 33; impending shortage of, 130; I.Q.—test results, 34
Physics, 77, 82; defined, 88; historic roots of, 78; history of, 88-89; interdisciplinary specialties, chart 100-101; recent growth, 123-124; specialties, chart 88-89, 100
Physiology, 96, 100, 101, 150
Picard, Jean, 95
Picasso, Pablo, 167
Pilkington, Alastair, 11
Planck, Max, 88
Planetary motion, Kepler's laws, 53, 81
Plants, classification of, 82, 97
Plasma physics, 88, 89, 100
Plato, 52, 59, 79, 98, 104
Poetry: computer, 173; impact of science on, 166
Point Arguello, Calif., map 131
Political repression of scientists, 35-36, 112
Political science, 98, 101
Polling, opinion, 150, 170
Polymer chemistry, 91, 100
Polymerization, 90, 91
Popularization of science, 103, 106-107
Population figures, scientific community vs. general growth, 124, 129-130
Positional astronomy, 92, 101
Pragmatic science, 77; history of, 77-78, 79; present role of, 30
Predictive sciences, 76-77
President's Panel on International Science, 120
Prestige of scientists, 34, 36
Prince, The, Machiavelli, 98
Princeton, N.J., Institute for Advanced Study, 127
Principia Mathematica, Newton, 89
Printing: early illustration, 81; invention of, 81; mechanization and automation of, 151
Probability and statistics, mathematical discipline, 87, 100
Probability chart, 61
Probability theory, 57, 61, 86, 172-173
Protocol facts, 54-55, diagram 58
Protocol plane, 54
Proton, 64, 66
Psychologist(s): dating of term, 29; I.Q.—test results, 34
Psychology, 76, 98, 99, 101
Ptolemy, 79, 92, 104
Publishing, science, 106-108, graph 109
Pythagoras, 79, 104

Q

Quadrant, 38, 39
Qualitative fact, 55
Quantitative fact, 55
Quantum chemistry, 91, 100
Quantum mechanics, 88, 89, 100
Quantum number rating of atomic particles, 64, 65
Quantum physics, 61, 172
Quantum statistics, theory of, 110

R

Radar, 147, 152
Radio: inventions leading to, chart 146-147; period of development, 148, graph 149
Radio astronomy, 93, 100, 101
Radio telescope, 92, 93, 116, 128
Radioactive dating, 165
Radiobiology, 97, 100, 101
Railroads: initial public reaction, 150, 155; electrification, 156
Ramsden, Jesse, 43
Rationalism vs. empiricism, 59
Rayleigh, Lord, 186
Reality: elusiveness, in advanced science, 58-62; nature of, philosophic debate, 59
Refrigeration, 146
Relativity, lithograph, 164
Relativity, theory of, 88, 89, 145; curved path of light, diagram 57
Relativity physics, 89, 100
Religion: effects of science on, 94, 165, 170-173; nature of, 171; of scientists, 31, 33-34
Renaissance, 103, 104; contributions to science, 77, 81; instrument-making, 36, 42-43; perspective in painting, 166
Republic, The, Plato, 98
Research, applied, financing, graph 124
Research, basic: federal programs and agencies, 127; financing, graph 124, 127-129; institutes, 128-129, 139; and technology, 145-147; U.S. a latecomer in, 146-147; U.S. expenditure for, 124, 130, 141
Research equipment, cost of, 128, 141
Research scientists, 30, 31; number of, in U.S., graph 31; personality pattern, 32-33
Riemann, Bernhard, 86, 87
Robida, Albert, 156-157
Robots, predictions for, 148; 1884 cartoon, 151
Rockefeller Foundation, 129
Roe, Anne, 32-33
Röntgen, Wilhelm Conrad, 184
Rosse, Earl of, 128
Rossi-Fanelli, Allessandro, 112
Royal Academy of Science, Sweden, 178-179
Royal Caroline Institute of Medicine, Sweden, 178, 179
Royal Society (of London), 122, 125, 126, 146
Rutherford, Ernest, 48

S

Sacramento, Calif., map 131
Salaries, scientists', 130, 142
Samios, Nicholas, 65, 66, 70-71
San Diego, Calif., map 131
San Francisco, as center of science, 130, map 131
Santa Barbara, Calif., map 131
Santos-Dumont, Alberto, 158
Sarabhai, Vikram, 117
Satellite systems, future, 149
Satellites, 1929 conception of, 163
Savannah, the, 144
Schleiden, Matthias, 83
Schulze, Johann Heinrich, 148
Schwann, Theodor, 83
Science: branches of, 75, 77, 84, 85-101 (see also Specialties); definitions, 29, 146; nature of, as distinct from other knowledge, 30, 75; need for cross-fertilization with humanities, 109-110, 173; origin of term, 29; ranking order within, 76; recent growth, and future growth trend, 123-124, 129-130
Science, la, French term, 29
Science, periodical, 107
Science Advisory Committee, 126-127
Science fiction, 160, 162, 173
Science teachers, number of, in U.S., graph 31, 124
Scientific method, 30, 51-58, 62, 77, 81, 145, 173; deduction of theory, 56-57; essence (steps) of, 52, diagram 58; measurement, 54-56; not equally workable in all science, 76, 82; observation, 54; verification of theory, 57-58
Scientist(s): dating of term, 29; definitions, 29; I.Q. level, 32; origin of term, 29; personality pattern, 32-33; popular image of, 30-31; prestige of, 34, 36; range of meaning of term, 30; shortage of, 130; types of mentality, 30. See also Statistics on science and scientists
Scripps Institution of Oceanography, 141
Senarens, Lu, 156
Severinus, 82
Sextants, 38-39
Shannon, Claude, 147
Shapley, Harlow, 172
Shaw, George Bernard, 173
Shawlow, Arthur, 136
Shelley, Percy Bysshe, 166
Shostakovich, Dimitri, 166
Shutt, Ralph P., 66
Silicon crystal, 24-25
Silliman, Benjamin, 106
Simpson, George Gaylord, 32, 107
Smith, Adam, 82, 99
Snow, C. P., 104
Social sciences, 75, 77, 147; beginnings, 82; history of, 98-99; interdisciplinary specialties, chart 100-101; specialties, chart 98-99, 101
Social scientists, 30; number in U.S., 124; personality, 33; divorce rate, 33
Societies, scientific, 125-126
Sociology, 76, 98, 99, 101

Solar battery, 148
Solid-state physics, 88, 89, 100, 149
Space exploration, 92
Space fiction, 162
Spatial perception, scientists', 32; test problems and answers, 32, 34
Specialties, 75-77, 82, 84, 85, charts 86-101; dating of terms for, 29; interdisciplinary, 75, chart 100-101; number of, 75; problems of communication between, 107
Spectroscope, 92
Spectroscopy, stellar, 93, 116
Spencer, Herbert, 16
Sputnik, impact of, 106, 126
Stanford, Calif., Center for Advanced Study in the Behavioral Sciences, 127
Stanford Industrial Park, 136
Stanford Research Institute, 136, 137
Stanford University, 136-137, 141; linear accelerator, 140-141
Starlight, bending of (theory of relativity), diagram 57
Static-electricity generator, 45
Statistical analysis, 84
Statistical mechanics, 89, 100
Statistics, mathematical discipline, 87, 100
Statistics on science and scientists: cost of research equipment, 128; distribution of scientists in U.S., map 30; divorce rates of scientists, 33; groups within U.S. scientific community, graph 31, 124; growth of science, 124, 129-130; home background of scientists (religious, class, regional), 33-34; scientists' I.Q. levels, 32; number of living scientists, 30; Ph.Ds and academicians, 30; total number of U.S. scientists, 124; U.S. expenditures for science, 124
Steam engine, development of, 146, 147-148, 154-155; Hero, 88, 147
Stewart, Walter, 10
Stockhausen, Karlheinz, 168
Stonehenge, 78
Stratigraphy, 95, 101
Stravinsky, Igor, 169
Strong, Foster, 139
Strong force, nuclear, 64, 65, 72
Structural geology, 95, 101
Subatomic particles. See Particles
Submarines, 150, 160
Sullivan, Louis, 167
Sumerians, contributions to science, 78
Sundials, 42, 54
Swings, Pol, 116
Synchrocyclotron, Berkeley, 135
Synchrotrons, 152; Brookhaven, diagram 66, 128
Synthetic compounds, 90, 91
Systems, complex, 83-84
Szilard, Leo, 9

T

Tartaglia, Niccolò, 86
Taxonomy, 97, 101
Taylor, Sir Geoffrey, 11
Teachers, science, number in U.S., graph 31, 124
Teamwork, scientific, 113, 123
Technicians, number in U.S., graph 31, 124
Technology, 145-148; benefits of, vs. damages and dangers, 145, 150-152; defined, 145; inevitability of advance, 152; predictions of past fulfilled, 148, 150-151, 152, 153; predictions for future developments, 148-150; sociological impact of, 145, 150-151, 155
Telegraph, 146
Telephone, 146, 147, 153; period of development, 148, graph 149; video, 149, 156
Telephonoscope, 156
Telescopes, 36, 92, 153; cost of, statistics, 128; invention of, 40; Herschel's, 93; modern, 116, 133; Newton's, 40; radio, 92, 93, 116, 128; types of, 40
Television, 148, 156; period of development, 148, graph 149; predictions for future, 149
Temperature scales, 56
Terman, Frederick E., 136
Terminology, scientific, 35, 105-106, examples 107
Testing of scientists, I.Q. and personality, 32-33; I.Q., problems and answers, 32, 34
Theoretical sciences, 76; and applied science, 145-148; beginnings of, 79-80
Theoretical scientists, 30, 51, 82
Theory, scientific: formulating, 57, diagram 58; interaction with fact, 53-58, diagram 58; verification of, 57-58
Thermodynamics, 76, 88, 89, 100, 146
Thomson, Sir George, 174
Thomson, J. J., 48
Time measurement, advances in, 54-55
Topology, 57, 87, 100

Transistors, period of development, 148, graph 149
Transportation, predictions for future innovations, 149
Tree trunk, cross section, 18-19
Trigonometry, 86, 100
Triode radio tube, 136
Twentieth Century, The, Robida, 156-157

U

Uncertainty principle, Heisenberg's, 61, 172-173
Underwater life, predictions for future, 150; 1910 cartoon, 161
Unge, Wilhelm, 176
Unified field theory, 72, 83
United States: distribution of scientists in, map 30; funds for science, sources of, graph 124, 141; Government role in sciences, 123, 125, 126-127, 128-129; a latecomer in basic research, 146-147; scientific community, groups within, graph 31, 124; scientific "establishment" in, 123, 124-129; total number of scientists, 124
Universe: expansion of, 58, 83; workings of, principle of uncertainty vs. determinism, 172-173
Universities: combined science-humanities curricula, 109; research centers, 127-129; science teaching, 30, 125, 142
University of Arizona, 129
University of California, 129, 130; Berkeley, 132-135; Los Angeles, 138-139; San Diego, 140-141
University of Chicago, 129
University of Wisconsin, 128
Urey, Harold, 140
U.S. Army Corps of Engineers, 126
U.S. Coast and Geodetic Survey, 126
U.S. Naval Observatory, 125
U.S. Patent Office, 126
U.S. Weather Bureau, 126
Utopia, More, 98

V

Vacuum tubes, inventions, 136, 146, 148
Van Allen, James A., 11
Varian, Russell, 136
Verne, Jules, 160
Verbal aptitude of scientists, 32, 34
Verbal vs. numerical definitions, 55
Verification, as step in scientific method, 52, 57-58
Videophone, 149, 156
Volcanology, 115
Volta, Alessandro, 89
Von Neumann, John, 13, 147

W

Walton, Ernest, 48
Watkins, Dean, 137
Watt, James, 80, 146, 147, 155
Wealth of Nations, The, Smith, 99
Wells, H.G., 106
Werner, Abraham Gottlob, 82
Whewell, William, 29, 82
Wiener, Norbert, 147
Wigner, Eugene P., 9, 179, 180-181
Wilson cloud chamber, 48
Wissenschaft, die, German term, 29
Wohler, Friedrich, 90
Women in science: 19th Century students, 127; Nobel laureates, 180
Woods Hole Oceanographic Institute, 128
Wren, Christopher, 11
Wright, Frank Lloyd, 167
Wright, Irving S., 50
Wright brothers, 150, 158
Written records, early, 78, 81

X

Xenakis, Iannis, 167-168

Y

Yale University, 109
Yalouris, Nikos, 114, 115
Yang, Chen Ning, 8, 187
Yeats, William Butler, 166
Yellow-pine tree trunk, cross section, 18-19
Youth, attitudes toward scientific career, 31
Yukawa, Hideki, 65

Z

Zoologist, dating of term, 29
Zoology, 96, 101

PICTURE CREDITS

The sources for the illustrations in this book are shown below. Credits for pictures from left to right are separated by commas, top to bottom by dashes.

Cover—Arnold Newman.

CHAPTER 1: 8—Brookhaven National Laboratories. 10—© 1955 *The New Yorker Magazine*, Inc. 11—© 1964 *The New Yorker Magazine*, Inc. 12—© 1954 *The New Yorker Magazine*, Inc. 14—© 1947 *The New Yorker Magazine*, Inc. 15—© 1955 *The New Yorker Magazine*, Inc. 17—Erwin W. Mueller, Pennsylvania State University. 18, 19—Rutherford Platt except right Dr. William M. Harlow from Photo Researchers, Inc. 20, 21—Dr. Roman Vishniac (2)—Eric V. Gravé from Photo Researchers, Inc., Stephen Collins from Photo Researchers, Inc. 22—General Motors Research Laboratories, Warren, Mich. 23—Leo Stashin from Rapho-Guillumette. 24, 25—General Motors Research Laboratories, Warren, Mich. except bottom center Jack Kath for Merck, Sharp & Dohme Research Laboratories. 26, 27—Leo C. Massopust.

CHAPTER 2: 28—Myron H. Davis. 30, 31—Drawings by Otto van Eersel. 32, 34—Drawings by Nicholas Fasciano based on *The Making of a Scientist* by Anne Roe, Dodd, Mead, copyright 1952, 1953 by Anne Roe. 35—The Bettmann Archive. 37—Walter Sanders courtesy Museo di Storia della Scienza, Florence. 38—Left A. G. Ingram Ltd. courtesy Royal Observatory, Edinburgh; right Walter Sanders, Stift Kremsmünster Oberösterreich. 39—William Abbenseth courtesy Rare Books Department, General Library, University of California, Berkeley. 40—Dr. Heinz Zinram courtesy Royal Society, London—left Eric Schaal courtesy Medical Museum of the Armed Forces, Institute of Pathology, Washington; right Derek Bayes courtesy Science Museum, London (2). 41—Roger Guillemot, *Connaissance des Arts*, from collection of Pierre G. Bernard, Paris. 42, 43—Roger Guillemot, *Connaissance des Arts*, from collection of Jacques Kugel, Paris; Dr. Heinz Zinram courtesy The Whipple Museum, Cambridge; Robert Lackenback from Black Star courtesy Zwinger Museum, Dresden—Dr. Heinz Zinram courtesy Museum of the History of Science, Oxford. 44—Derek Bayes courtesy Trustees of the Tate Gallery. —Robert Mottar courtesy University of Lund, Sweden—Eric Schaal courtesy Dr. Bern Dibner, Burndy Library, Norwalk, Connecticut. 46, 47—Walter Sanders courtesy Liebig Museum, Giessen. 48—Professor Derek de Solla Price—Cavendish Laboratory, Cambridge. 49—Cavendish Laboratory, Cambridge.

CHAPTER 3: 50—Carl Bakal. 54, 55—Drawings by Anthony Saris. 57—Drawing by Otto van Eersel. 58—Drawing by Bob Pellegrini. 61—Drawing by Charles Mikolaycak. 63—Brookhaven National Laboratories. 64, 65—J. R. Eyerman (2), Art Rickerby, J. R. Eyerman, Cern, Geneva, Switzerland (3)—drawings by Charles Mikolaycak. 66—Brookhaven National Laboratories—drawing by Adolph E. Brotman, Art Rickerby. 67—Drawing by George V. Kelvin. 68, 69—Art Rickerby. 70, 71—Top Art Rickerby; bottom Brookhaven National Laboratories except left Burk Uzzle. 72—Jim Mahan. 73—Burk Uzzle.

CHAPTER 4: 74—Phil Brodatz courtesy Burndy Library, Norwalk, Conn. 76—The Bettmann Archive—Phil Brodatz courtesy Burndy Library, Norwalk, Conn. 78—Drawings by Nicholas Fasciano. 81—The New York Public Library. 83—Drawing by Matt Greene. 85—Drawings by Leo and Diane Dillon. 86 through 99—Drawings by Leo and Diane Dillon, graphs by George V. Kelvin. 100, 101—Drawing by George V. Kelvin.

CHAPTER 5: 102—Eric Schaal. 104—The Newberry Library. 106—Reproduced with the kind permission of the Beinecke Rare Book and Manuscript Library, Yale University. 109—Drawings by Otto van Eersel. 111—Baldev from Pix. 112—© Philippe Halsman, Alfred Eisenstaedt, Robert Lackenback from Black Star. 113—Alfred Eisenstaedt, Francisco Vera. 114, 115—Alfred Eisenstaedt—Willimetz, Larry Burrows. 116—Pierre Boulat, David Miller. 117—Brian Seed, T. S. Satyan. 118—David Lees, Ettore Naldoni. 119—Larry Sherman from Black Star. 120—Phil Bath. 121—© Philippe Halsman.

CHAPTER 6: 122—Brian Seed. 124, 125—Drawings by Otto van Eersel. 127—The Bettmann Archive. 131—Background drawing by Jerome Snyder, photograph by Henry Groskinsky. 132, 133—Ralph Crane from Black Star, Lawrence Radiation Laboratory—Jon Brenneis. 134, 135—Jon Brenneis. 136, 137—Wayne Miller from Magnum except top left Stanford University. 138, 139—Phil Bath—Wayne Miller from Magnum. 140, 141—Jon Brenneis—Wayne Miller from Magnum. 142, 143—J. R. Eyerman.

CHAPTER 7: 144—The Babcock & Wilcox Company. 149—Drawings by Otto van Eersel. 150—The New York Public Library—The Bettmann Archive. 151—The Bettmann Archive. 153—Drawing by Lowell Hess. 154, 155—Right Pierre Belzeaux from Rapho-Guillumette courtesy Conservatoire National des Arts et Métiers. 156—Brown Brothers—Jean Marquis, The Bettmann Archive. 157—Jean Marquis. 158, 159—The New York Public Library—I. Sy Seidman. 160—Bottom I. Sy Seidman. 162—Phil Brodatz courtesy Sam Moscowitz, © originally by Hugo Gernsback, New York; © originally by Hugo Gernsback, New York. 163—Phil Brodatz courtesy Sam Moscowitz, © originally by Hugo Gernsback, New York.

CHAPTER 8: 164—Courtesy Mikelson's Gallery, Washington, D.C. 166—Courtesy Uffizi, Florence. 168—© 1956 by Universal Edition (London) Ltd., London. 173—Librascope Division of General Precision, Inc., Glendale, California. 175—Cornell Capa from Magnum. 176—Courtesy the Nobel Foundation. 177—Marvin E. Newman courtesy of Olaf Nobel. 178, 179—Marvin E. Newman courtesy of the Nobel Foundation, Brian Seed, Marvin E. Newman (2)—J. R. Eyerman. 180, 181—Tore Johnson (Tio) from Black Star. 182, 183—Tore Johnson (Tio) from Black Star except bottom left Lennart Nilsson. 184—Leo Gundermann, Novosti Press Agency—Jean Marquis, Alan Clifton, Jean Marquis, Alan Clifton. 185—Phil Brodatz courtesy the Einstein Estate. 186—Pressens Bild A. B., Stockholm—International Magazine Service. 187—A. B. Reportagebild, Stockholm.

APPENDIX: 189—United Press International, courtesy the Nobel Foundation (2), Sallstedts Bildbyra courtesy the Nobel Foundation, Wide World Photos—Sallstedts Bildbyra courtesy the Nobel Foundation, United Press International, Sallstedts Bildbyra courtesy the Nobel Foundation (2), courtesy the Nobel Foundation—courtesy the Nobel Foundation (6)—courtesy the Nobel Foundation, United Press International, courtesy the Nobel Foundation (2), Sallstedts Bildbyra courtesy the Nobel Foundation, Wide World Photos. 190—Sallstedts Bildbyra courtesy the Nobel Foundation (3), courtesy the Nobel Foundation, Freelance Photographers Guild—Freelance Photographers Guild, Sallstedts Bildbyra courtesy the Nobel Foundation (4), Keystone Press Agency, Sallstedts Bildbyra courtesy the Nobel Foundation (4), Keystone Press Agency, Sallstedts Bildbyra courtesy the Nobel Foundation—Sallstedts Bildbyra courtesy the Nobel Foundation (6)—Sallstedts Bildbyra courtesy the Nobel Foundation (3), Sovfoto, Sallstedts Bildbyra courtesy the Nobel Foundation (2)—James F. Coyne, Sallstedts Bildbyra courtesy the Nobel Foundation, Jon Brenneis, Wide World Photos, United Press International, Roy Jarvis—United Press International, courtesy the Nobel Foundation (2), Brown Brothers, courtesy the Nobel Foundation—Sallstedts Bildbyra courtesy the Nobel Foundation (4), courtesy the Nobel Foundation. 191—Sallstedts Bildbyra courtesy the Nobel Foundation (2), courtesy the Nobel Foundation (3)—courtesy the Nobel Foundation, Sallstedts Bildbyra courtesy the Nobel Foundation (2)—courtesy the Nobel Foundation, Brown Brothers, Sallstedts Bildbyra courtesy the Nobel Foundation, courtesy the Nobel Foundation, Brown Brothers—courtesy the Nobel Foundation (4), Brown Brothers, courtesy the Nobel Foundation—Wide World Photos, Sallstedts Bildbyra courtesy the Nobel Foundation (2), Brown Brothers, Wide World Photos, Sallstedts Bildbyra courtesy the Nobel Foundation—Alan W. Richards, Sallstedts Bildbyra courtesy the Nobel Foundation, Sovfoto, Sallstedts Bildbyra courtesy the Nobel Foundation (2)—courtesy the Nobel Foundation (6)—courtesy the Nobel Foundation, Brown Brothers, Sallstedts Bildbyra courtesy the Nobel Foundation, The Bettmann Archive, Sallstedts Bildbyra courtesy the Nobel Foundation (2). 192—Brown Brothers, courtesy the Nobel Foundation (4)—courtesy the Nobel Foundation, Sallstedts Bildbyra courtesy the Nobel Foundation (3), Fritz Eschen, Tita Binz—Farabola, Milan; courtesy the Nobel Foundation (3); New York Academy of Medicine—courtesy the Nobel Foundation, Sallstedts Bildbyra courtesy the Nobel Foundation (5)—Sallstedts Bildbyra courtesy the Nobel Foundation (6)—Fritz Eschen (2), Sallstedts Bildbyra courtesy the Nobel Foundation (2), Wide World Photos, Sallstedts Bildbyra courtesy the Nobel Foundation—Sallstedts Bildbyra courtesy the Nobel Foundation (6)—Sallstedts Bildbyra courtesy the Nobel Foundation (6). 193—Sallstedts Bildbyra courtesy the Nobel Foundation (4), Wide World Photos, Roma's Press Photo—Fritz Eschen, Sallstedts Bildbyra courtesy the Nobel Foundation (3), Robert Lackenback, Sallstedts Bildbyra courtesy the Nobel Foundation—Sallstedts Bildbyra courtesy the Nobel Foundation, David Gahr, Sallstedts Bildbyra courtesy the Nobel Foundation, Wide World Photos, Sallstedts Bildbyra courtesy the Nobel Foundation—© Fabian Bachrach, J. R. Eyerman, Sallstedts Bildbyra courtesy the Nobel Foundation (2), courtesy the Nobel Foundation—courtesy the Nobel Foundation (5), Sallstedts Bildbyra courtesy the Nobel Foundation—Sallstedts Bildbyra courtesy the Nobel Foundation (2), Sy Friedman, courtesy the Nobel Foundation (2)—courtesy the Nobel Foundation, Sallstedts Bildbyra courtesy the Nobel Foundation (5)—Sallstedts Bildbyra courtesy the Nobel Foundation (3), Keystone Press Agency, United Press International (2). 194—Courtesy the Nobel Foundation (5)—courtesy the Nobel Foundation except fourth from left—courtesy the Nobel Foundation, The Bettmann Archive, Sallstedts Bildbyra courtesy the Nobel Foundation (4)—Tommy Weber, Wide World Photos, Sallstedts Bildbyra courtesy the Nobel Foundation (2), United Press International, Sallstedts Bildbyra courtesy the Nobel Foundation—Sallstedts Bildbyra courtesy the Nobel Foundation (4), Columbia University, Sallstedts Bildbyra courtesy the Nobel Foundation—Sallstedts Bildbyra courtesy the Nobel Foundation (2), Camera Press from Pix. Phil Brodatz courtesy the Einstein Estate. Back cover—Charles Mikolaycak.

PRODUCTION STAFF FOR TIME INCORPORATED

Arthur R. Murphy Jr. (Vice President and Director of Production), Robert E. Foy, James P. Menton, Caroline Ferri and Robert E. Fraser
Text photocomposed under the direction of Albert J. Dunn and Arthur J. Dunn

Printed by R. R. Donnelley & Sons Company, Crawfordsville, Indiana,
and Livermore and Knight Co., a division of Printing Corporation of America, Providence, Rhode Island
Bound by R. R. Donnelley & Sons Company, Crawfordsville, Indiana
Paper by The Mead Corporation, Dayton, Ohio
Cover stock by The Plastic Coating Corporation, Holyoke, Massachusetts